COMPUTATIONAL LIQUID CRYSTAL PHOTONICS

COMPUTATIONAL LIQUID CRYSTAL PHOTONICS

FUNDAMENTALS, MODELLING AND APPLICATIONS

Salah Obayya
Center for Photonics and Smart Materials
Zewail City of Science and Technology
Giza
Egypt

Mohamed Farhat O. Hameed and Nihal F.F. Areed
Center for Photonics and Smart Materials
Zewail City of Science and Technology
Giza
and
Faculty of Engineering
Mansoura University
Mansoura
Egypt

This edition first published 2016
© 2016 John Wiley & Sons, Ltd.

Registered Office
John Wiley & Sons, Ltd, The Atrium, Southern Gate, Chichester, West Sussex, PO19 8SQ, United Kingdom

For details of our global editorial offices, for customer services and for information about how to apply for permission to reuse the copyright material in this book please see our website at www.wiley.com.

Library of Congress Cataloging-in-Publication data applied for

ISBN: 9781119041955

A catalogue record for this book is available from the British Library.

Set in 10/12pt Times by SPi Global, Pondicherry, India
Printed and bound in Singapore by Markono Print Media Pte Ltd

1   2016

All Praise is due to Allah, and peace and blessings be upon. Prophet Muhammad and upon his family and his Companions.

The authors would like to dedicate this book to Prof. Ahmed Zewail for his continuous encouragement, support, and the opportunity to contribute to the Egypt National project of renaissance: Zewail City of Science and Technology.

The authors would also like to dedicate the book to their families, whose love, support, patience, and understanding are beyond any scope.

Contents

Preface

The turn toward optical computers and photonic integrated circuits in high-capacity optical networks has attracted the interest of expert researchers. This is because all optical packet switching and routing technologies can provide more efficient power and footprint scaling with increased router capacity. Therefore, it is aimed to integrate more optical processing elements into the same chip and, hence, on-chip processing capability and system intelligence can be increased. The merging of components and functionalities decreases packaging cost and can bring photonic devices one step (or more) closer to deployment in routing systems.

Photonic crystal devices can be used functionally as part of a comprehensive all-photonic crystal-based system where, on the same photonic crystal platform, many functionalities can be realized. Therefore, photonic crystals have recently received much attention due to their unique properties in controlling the propagation of light. Many potential applications of photonic crystals require some capability for tuning through external stimuli. It is anticipated that photonic crystals infiltrated with liquid crystals (LCs) will have high tunability with an external electric field and temperature. For the vast majority of LCs, the application of an electric field results in an orientation of the nematic director either parallel or perpendicular to the field, depending on the sign of the dielectric anisotropy of the nematic medium. The scope of this book is to propose, optimize, and simulate new designs for tunable broadband photonic devices with enhanced high levels of flexible integration and enhanced power processing, using a combination of photonic crystal and nematic LC (NLC) layers. The suggested NLC photonic devices include a coupler, a polarization splitter, a polarization rotator, and a multiplexer–demultiplexer for telecommunication applications. In addition, LC photonic crystal-based encryption and decryption devices will be introduced and LC-based routers and sensors will be presented. In almost all cases, an accurate quantitative theoretical modeling of these devices has to be based on advanced computational techniques that solve the corresponding, numerically very large linear, nonlinear, or coupled partial differential equations. In this regard, the book will also offer an easy-to-understand, and yet comprehensive, state-of-the-art of computational modeling techniques for the analysis of lightwave propagation in a wide range of LC-based modern photonic devices.

There are many excellent books on LCs; however, several of these concentrate on the physics and chemistry of the LCs especially for LC display (LCD) applications. In addition, many books on photonic devices have been published in the recent years. However, it is still difficult to find one book in which highly tunable photonic crystal devices based on LC materials are discussed with a good balance of breadth and depth of coverage. Therefore, the book will represent a unique source for the reader to learn in depth about the modeling techniques and simulation of the processing light through many tunable LC devices.

The primary audience for this book are undergraduate students; the student will be taken from scratch until he can develop the subject himself. The secondary audience are the business and industry experts working in the fields of information and communications technology, security, and sensors because the book intends to open up new possibilities for marketing new commercial products. The audience of this book will also include the researchers at the early and intermediate stages working in the general areas of LC photonics. The book consists of three parts: LC basic principles, numerical modeling techniques, and LC-based applications. The first part includes three chapters where the basic principles of waveguides and modes, photonic crystals, and liquid crystals are given. From Chapters 4 to 6, the numerical techniques operating in the frequency domain are presented. Among them, Chapter 4 presents the governing equations for the full-vectorial finite-difference method (FVFDM) and perfectly matched layer (PML) scheme for the treatment of boundary conditions. The FVFDM is then assessed in Chapter 5 where the modal analysis of LC-based photonic crystal fiber (PCF) is given. The FV beam propagation method (FVBPM) is presented in Chapter 6 to study the propagation along the LC-PCF-based applications. After deriving the governing equations, the FVBPM is numerically assessed through several optical waveguide examples. The conventional finite-difference time domain (FDTD) method in 2D and 3D, as an example of the numerical techniques operating in the time domain is presented in Chapter 7.

The third part consists of six chapters to cover the applications of the LC-based photonic crystal devices. From Chapters 8 to 10, the applications of the LC-PCF for telecommunication devices, such as couplers, polarization rotators, polarization splitters, and multiplexer–demultiplexers, are introduced. In addition, the LC-PCF sensors, such as biomedical and temperature sensors, are explained in Chapter 11. Photonic crystal-based encryption systems for security applications are covered in Chapter 12. Optical computing devices, such as optical routers, optical memory, and reconfigurable logic gates, are introduced in Chapter 13.

Part I

Basic Principles

Part I

Basic Principles

1

Principles of Waveguides

1.1 Introduction

A waveguide can be defined as a structure that guides waves, such as electromagnetic or sound waves [1]. In this chapter, the basic principles of the optical waveguide will be introduced. Optical waveguides can confine and transmit light over different distances, ranging from tens or hundreds of micrometers in integrated photonics, to hundreds or thousands of kilometers in long-distance fiber-optic transmission. Additionally, optical waveguides can be used as passive and active devices such as waveguide couplers, polarization rotators, optical routers, and modulators. There are different types of optical waveguides such as slab waveguides, channel waveguides, optical fibers, and photonic crystal waveguides. The slab waveguides can confine energy to travel only in one dimension, while the light can be confined in two dimensions using optical fiber or channel waveguides. Therefore, the propagation losses will be small compared to wave propagation in open space. Optical waveguides usually consist of high index dielectric material surrounded by lower index material, hence, the optical waves are guided through the high index material by a total internal reflection mechanism. Additionally, photonic crystal waveguides can guide the light through low index defects by a photonic bandgap guiding technique. Generally, the width of a waveguide should have the same order of magnitude as the wavelength of the guided wave.

In this chapter, the basic optical waveguides are discussed including waveguides operation, Maxwell's equations, the wave equation and its solutions, boundary conditions, phase and group velocity, and the properties of modes.

Computational Liquid Crystal Photonics: Fundamentals, Modelling and Applications, First Edition.
Salah Obayya, Mohamed Farhat O. Hameed and Nihal F.F. Areed.
© 2016 John Wiley & Sons, Ltd. Published 2016 by John Wiley & Sons, Ltd.

1.2   Basic Optical Waveguides

Optical waveguides can be classified according to their geometry, mode structure, refractive index distribution, materials, and the number of dimensions in which light is confined [2]. According to their geometry, they can be categorized by three basic structures: planar, rectangular channel, and cylindrical channel as shown in Figure 1.1. Common optical waveguides can also be classified based on mode structure as single mode and multiple modes. Figure 1.1a shows that the planar waveguide consists of a core that must have a refractive index higher than the refractive indices of the upper medium called the cover, and the lower medium called the substrate. The trapping of light within the core is achieved by total internal reflection. Figure 1.1b shows the channel waveguide which represents the best choice for fabricating integrated photonic devices. This waveguide consists of a rectangular channel that is sandwiched between an underlying planar substrate and the upper medium, which is usually air. To trap the light within a rectangular channel, it is necessary for the channel to have a refractive index greater than that of the substrate. Figure 1.1c shows the geometry of the cylindrical channel waveguide which consists of a central region, referred to as the core, and surrounding material called cladding. Of course, to confine the light within the core, the core must have a higher refractive index than that of the cladding.

Figure 1.2 shows the three most common types of channel waveguide structures which are called strip, rip, and buried waveguides. It is evident from the figure that the main difference between the three types is in the shape and the size of the film deposited onto the substrate. In the strip waveguide shown in Figure 1.2a, a high index film is directly deposited on the substrate with finite width. On the other hand, the rip waveguide is formed by depositing a high index film onto the substrate and performing an incomplete etching around a finite width as shown in Figure 1.2b. Alternatively, in the case of the buried waveguide shown in Figure 1.2c,

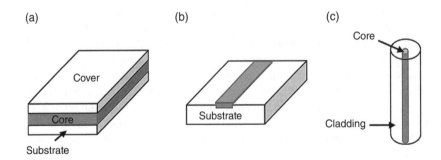

Figure 1.1   Common waveguide geometries: (a) planar, (b) rectangular, and (c) cylindrical

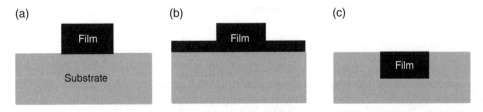

Figure 1.2   Common channel waveguides: (a) strip, (b) rip, and (c) buried

diffusion methods [2] are employed in order to increase the refractive index of a certain zone of the substrate.

Figure 1.3 shows the classification of optical waveguides based on the number of dimensions in which the light rays are confined. In planar waveguides, the confinement of light takes place in a single direction and so the propagating light will diffract in the plane of the core. In contrast, in the case of channel waveguides, shown in Figure 1.3b, the confinement of light takes place in two directions and thus diffraction is avoided, forcing the light propagation to occur only along the main axis of the structure. There also exist structures that are often called photonic crystals that confine light in three dimensions as revealed from Figure 1.3c. Of course, the light confinement in this case is based on Bragg reflection. Photonic crystals have very interesting properties, and their use has been proposed in several devices, such as waveguide bends, drop filters, couplers, and resonators [3].

Classification of optical waveguides according to the materials and refractive index distributions results in various optical waveguide structures, such as step index fiber, graded index fiber, glass waveguide, and semiconductor waveguides. Figure 1.4a shows the simplest form of step index waveguide that is formed by a homogenous cylindrical core with constant

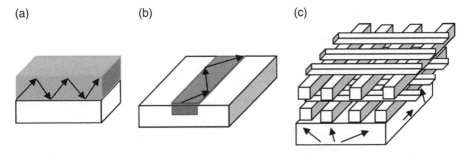

Figure 1.3 Common waveguide geometries based on light confinement: (a) planar waveguide, (b) rectangular channel waveguide, and (c) photonic crystals

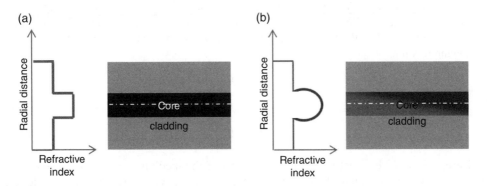

Figure 1.4 Classification of optical waveguide based on the refractive index distributions: (a) step-index optical fiber and (b) graded-index optical fiber

Table 1.1   The differential and integral forms of Maxwell's equations

Differential form	Integral form	
$\nabla \times E = \dfrac{-\partial B}{\partial t}$	$\displaystyle\oint_{\text{loop}} E \cdot dl = -\dfrac{\partial}{\partial t} \oint_{\text{area}} B \cdot dS$	(1.1)
$\nabla \times H = J + \dfrac{\partial D}{\partial t}$	$\displaystyle\oint_{\text{loop}} H \cdot dl = \int_{\text{area}} J \cdot dS + \dfrac{\partial}{\partial t} \int_{\text{area}} D \cdot dS$	(1.2)
$\nabla \cdot B = 0$	$\displaystyle\int_{\text{surface}} B \cdot dS = 0$	(1.3)
$\nabla \cdot D = \rho$	$\displaystyle\int_{\text{surface}} D \cdot dS = Q_{\text{enclosed}}$	(1.4)

refractive index surround by cylindrical cladding of a different, lower index. Figure 1.4b shows the graded index planar waveguide where the refractive index of the core varies as a function of the radial distance [4].

1.3   Maxwell's Equations

Maxwell's equations are used to describe the electric and magnetic fields produced from varying distributions of electric charges and currents. In addition, they can explain the variation of the electric and magnetic fields with time. There are four Maxwell's equations for the electric and magnetic field formulations. Two describe the variation of the fields in space due to sources as introduced by Gauss's law and Gauss's law for magnetism, and the other two explain the circulation of the fields around their respective sources. In this regard, the magnetic field moves around electric currents and time varying electric fields as described by Ampère's law as well as Maxwell's addition. On the other hand, the electric field circulates around time varying magnetic fields as described by Faraday's law. Maxwell's equations can be represented in differential or integral form as shown in Table 1.1. The integral forms of the curl equations can be derived from the differential forms by application of Stokes' theorem.

where E is the electric field amplitude (V/m), H is the magnetic field amplitude (A/m), D is the electric flux density (C/m^2), B is the magnetic flux density (T), J is the current density (A/m^2), ρ is the charge density (C/m^3), and Q is the charge (C). It is worth noting that the flux densities, D and B, are related to the field amplitudes E and H for linear and isotropic media by the following relations:

$$B = \mu H. \tag{1.5}$$

$$D = \varepsilon E. \tag{1.6}$$

$$J = \sigma E. \tag{1.7}$$

Here, $\varepsilon = \varepsilon_o \varepsilon_r$ is the electric permittivity (F/m) of the medium, $\mu = \mu_o \mu_r$ is the magnetic permeability of the medium (H/m), σ is the electric conductivity, ε_r is the relative dielectric constant, $\varepsilon_o = 8.854 \times 10^{-12}$ F/m is the permittivity of free space, and $\mu_o = 4\pi \times 10^{-7}$ H/m is the permeability of free space.

1.4   The Wave Equation and Its Solutions

The electromagnetic wave equation can be derived from Maxwell's equations [2]. Assuming that we have a source free ($\rho=0$, $J=0$), linear (ε and μ are independent of E and H), and isotropic medium. This can be obtained at high frequencies ($f > 10^{13}$ Hz) where the electromagnetic energy does not originate from free charge and current. However, the optical energy is produced from electric or magnetic dipoles formed by atoms and molecules undergoing transitions. These sources are included in Maxwell's equations by the bulk permeability and permittivity constants. Therefore, Maxwell's equations can be rewritten in the following forms:

$$\nabla \times E = \frac{-\partial B}{\partial t} \tag{1.8}$$

$$\nabla \times H = \frac{\partial D}{\partial t} \tag{1.9}$$

$$\nabla \cdot B = 0 \tag{1.10}$$

$$\nabla \cdot D = 0 \tag{1.11}$$

The resultant four equations can completely describe the electromagnetic field in time and position. It is revealed from Eqs. (1.8) and (1.9) that Maxwell's equations are coupled with first-order differential equations. Therefore, it is difficult to apply these equations when solving boundary-value problems. This problem can be solved by decoupling the first-order equations, and hence the wave equation can be obtained. The wave equation is a second-order differential equation which is useful for solving waveguide problems. To decouple Eqs. (1.8) and (1.9), the curl of both sides of Eq. (1.8) is taken as follows:

$$\nabla \times \left(\nabla \times E\right) = \nabla \times \frac{-\partial B}{\partial t} = \nabla \times \frac{-\partial \mu H}{\partial t} \tag{1.12}$$

If $\mu(r, t)$ is independent of time and position, Eq. (1.12) becomes thus:

$$\nabla \times \left(\nabla \times E\right) = -\mu \left(\nabla \times \frac{\partial H}{\partial t}\right) \tag{1.13}$$

Since the functions are continuous, Eq. (1.13) can be rewritten as follows:

$$\nabla \times \left(\nabla \times E\right) = -\mu \frac{\partial}{\partial t}\left(\nabla \times H\right) \tag{1.14}$$

Substituting Eq. (1.9) into Eq. (1.14) and assuming ε is time invariant, we obtain the following relation:

$$\nabla \times \left(\nabla \times E\right) = -\mu \frac{\partial}{\partial t}\left(\frac{\partial D}{\partial t}\right) = -\mu\varepsilon \frac{\partial^2 E}{\partial t^2} \tag{1.15}$$

The resultant equation is a second-order differential equation with $(\nabla \times \nabla \times)$ operator and with only the electric field E as one variable. By applying the vector identity,

$$\nabla \times \nabla \times E = \nabla(\nabla \cdot E) - \nabla^2 E, \tag{1.16}$$

where the ∇^2 operator in Eq. (1.16) is the vector Laplacian operator that acts on the E vector [2]. The vector Laplacian can be written in terms of the scalar Laplacian for a rectangular coordinate system, as given by

$$\nabla^2 E = \nabla^2 E_x \, \hat{x} + \nabla^2 E_y \, \hat{y} + \nabla^2 E_z \, \hat{z}, \tag{1.17}$$

where $\hat{x}$, $\hat{y}$, and $\hat{z}$ are the unit vectors along the three axes. Additionally, the scalar ∇^2's on the right-hand side of Eq. (1.17) can be expressed in Cartesian coordinates:

$$\nabla^2 = \frac{\partial}{\partial x^2} + \frac{\partial}{\partial y^2} + \frac{\partial}{\partial z^2}. \tag{1.18}$$

In order to obtain $\nabla \cdot E$, Eq. (1.11) can be used as follows:

$$\nabla \cdot \varepsilon E = \nabla \varepsilon \cdot E + \varepsilon \nabla \cdot E = 0 \tag{1.19}$$

As a result, $\nabla \cdot E$ can be obtained as follows:

$$\nabla \cdot E = -E \cdot \frac{\nabla \varepsilon}{\varepsilon} \tag{1.20}$$

Substituting Eqs. (1.16) and (1.20) into Eq. (1.15) results in

$$\nabla^2 E - \mu \varepsilon \frac{\partial^2 E}{\partial t^2} = -\nabla\left(E \cdot \frac{\nabla \varepsilon}{\varepsilon}\right) \tag{1.21}$$

If there is no gradient in the permittivity of the medium, the right-hand side of Eq. (1.21) will be zero. Actually, for most waveguides, this term is very small and can be neglected, simplifying Eq. (1.21) to

$$\nabla^2 E - \mu \varepsilon \frac{\partial^2 E}{\partial t^2} = 0 \tag{1.22}$$

Equation (1.22) is the time-dependent vector Helmholtz equation or simply the wave equation. A similar wave equation can be obtained as a function of the magnetic field by starting from Eq. (1.9).

$$\nabla^2 H - \mu \varepsilon \frac{\partial^2 H}{\partial t^2} = 0 \tag{1.23}$$

Equations (1.22) and (1.23) are the equations of propagation of electromagnetic waves through the medium with velocity u:

$$u = \frac{1}{\sqrt{\mu \varepsilon}} \tag{1.24}$$

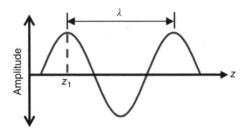

Figure 1.5 A traveling wave along the z-axis

It is worth noting that each of the electric and magnetic field vectors in Eqs. (1.23) and (1.24) has three scalar components. Consequently, six scalar equations for E_x, E_y, E_z, H_x, H_y, and H_z can be obtained. Therefore, the scalar wave equation can be rewritten as follows:

$$\nabla^2 \Psi - \frac{1}{u^2}\frac{\partial^2 \Psi}{\partial t^2} = 0 \tag{1.25}$$

Here, Ψ is one of the orthogonal components of the wave equations. The separation of variables technique can be used to have a valid solution [2].

$$\Psi(r,t) = \Psi(r)\phi(t) = \Psi_o \exp(jk \cdot r)\exp(j\omega t) \tag{1.26}$$

Here, Ψ_o is the amplitude, k is the separation constant which is well known as the wave vector (rad/m), and ω is the angular frequency of the wave (rad/s). The wave vector k will be used as a primary variable in most waveguide calculations. The magnitude of the wave vector that points in the propagation direction of a plane wave can be expressed as follows:

$$k = \omega\sqrt{\mu\varepsilon} \tag{1.27}$$

If the wave propagates along the z-axis, the propagation direction can be in the forward direction along the $+z$-axis with exponential term exp $(j\omega t - jkz)$ [2]. However, the propagation will be in the backward direction with exp$(j\omega t + jkz)$. Figure 1.5 shows the real part of the spatial component of a plane wave traveling in the z direction, $\Psi(z) = \Psi_o \exp(jkz)$. The amplitude of the wave at the first peak and the adjacent peak separated by a wavelength are equal, such that

$$e^{jkz_1} = e^{jk(z_1 + \lambda)} = e^{jkz_1}e^{jk\lambda} \tag{1.28}$$

Therefore, $e^{jk\lambda} = 1$, and hence $k\lambda = 2\pi$ which results in

$$k = \frac{2\pi}{\lambda} \tag{1.29}$$

1.5 Boundary Conditions

The waveguide in which the light is propagated is usually characterized by its conductivity σ, permittivity ε, and permeability μ. If these parameters are independent of direction, the material will be isotropic; otherwise it will be anisotropic. Additionally, the material is

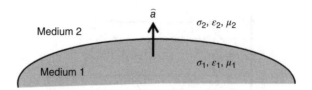

Figure 1.6 Interface between two mediums

homogeneous if σ, ε, and μ are not functions of space variables; otherwise, it is inhomogeneous. Further, the waveguide is linear if σ, ε, and μ are not affected by the electric and magnetic fields; otherwise, it is nonlinear. The electromagnetic wave usually propagates through the high index material surrounded by the lower index one. Therefore, the boundary conditions between the two media should be taken into consideration. Figure 1.6 shows the interface between two different materials 1 and 2 with the corresponding characteristics (σ_1, ε_1, μ_1) and (σ_2, ε_2, μ_2), respectively. The following boundary conditions at the interface [3] can be obtained from the integral form of Maxwell's equations with no sources, (ρ, $J=0$):

$$\hat{a} \times \left(E_{2t} - E_{1t} \right) = 0 \tag{1.30}$$

$$\hat{a} \times \left(H_{2t} - H_{1t} \right) = 0 \tag{1.31}$$

$$\hat{a} \cdot \left(B_{2n} - B_{1n} \right) = 0 \tag{1.32}$$

$$\hat{a} \cdot \left(D_{2n} - D_{1n} \right) = 0 \tag{1.33}$$

Here, $\hat{a}$ is a unit vector normal to the interface between medium 1 and medium 2, and subscripts t and n refer to tangent and normal components of the fields. It is revealed from Eqs. (1.30) and (1.31) that the tangential components of E and H are continuous across the boundary. In addition, the normal components of B and D are continuous through the interface, as shown in Eqs. (1.32) and (1.33), respectively.

1.6 Phase and Group Velocity

1.6.1 Phase Velocity

The propagation velocity of the electromagnetic waves is characterized by the phase velocity and the group velocity. Consider a traveling sinusoidal electromagnetic wave in the z direction. A point on one crest of the wave with specific phase must move at specific velocity to stay on the crest such that [5]

$$e^{-j(kz - \omega t)} = \text{constant} \tag{1.34}$$

This can be obtained if $(kz - \omega t) = \text{constant}$, and hence $z(t)$ must satisfy the following:

$$z(t) = \frac{\omega t}{k} + \text{constant} \tag{1.35}$$

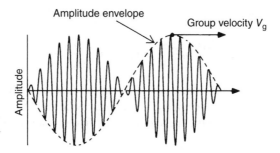

Figure 1.7   The superposition of two waves of different frequencies

The phase velocity, $v(t)=v_p$ can be obtained by differentiating $z(t)$ with respect to time as follows:

$$\frac{dz}{dt}=\frac{\omega}{k}=v_p \tag{1.36}$$

Therefore, the phase velocity v_p is a function of the angular frequency $\left(\omega = k/\sqrt{\mu\varepsilon}\right)$ and the magnitude of the wave vector. Then, the phase velocity can be rewritten in the following form:

$$v_p = \frac{1}{\sqrt{\mu\varepsilon}} \tag{1.37}$$

1.6.2   Group Velocity

The group velocity v_g [5] is used to describe the propagation speed of a light pulse. The group velocity can be expressed by studying the superposition of two waves of equal amplitude E_o but with different frequencies $\omega_1 = \omega + \Delta\omega$, and $\omega_2 = \omega - \Delta\omega$. Additionally, the corresponding wave vectors will be $k_1 = k + \Delta k$, $k_2 = k - \Delta k$, respectively. The superposition between the two waves can be expressed as follows:

$$E_t = E_1 + E_2 = E_o\left(\cos\left[(\omega+\Delta\omega)t-(k+\Delta k)z\right]+\cos\left[(\omega-\Delta\omega)t-(k-\Delta k)z\right]\right) \tag{1.38}$$

The resultant electric field can be rewritten as follows:

$$E_t = 2E_o\cos(\omega t - kz)\cos(\Delta\omega t - \Delta kz). \tag{1.39}$$

Therefore, a temporal beat at frequency $\Delta\omega$ and a spatial beat with period Δk is obtained, as shown in Figure 1.7. The envelope of the amplitude [2, 5] can be described by the $\cos(\Delta\omega t - \Delta kz)$ term and has a velocity equal to group the velocity v_g.

The group velocity can be obtained using the same procedure used in the calculation of the phase velocity. A point attached to the crest of the envelope, should move with a given speed to stay on the crest of the envelope, and hence the phase $(\Delta\omega t - \Delta kz)$ is constant.

Therefore, $z(t)$ can be expressed as follows:

$$z(t)=\frac{\Delta\omega t}{\Delta k}+\text{constant} \tag{1.40}$$

By applying the derivative of $z(t)$, the group velocity can be obtained as [5]:

$$v_g = \frac{dz(t)}{dt} = \frac{\Delta\omega}{\Delta k} \Rightarrow v_g = \lim_{\Delta\omega \to 0} \frac{\Delta\omega}{\Delta k} = \frac{d\omega}{dk}. \tag{1.41}$$

This means that the group velocity, v_g, is based on the first derivative of the angular frequency ω with respect to the wave vector k. Since $k = \omega n/c$, where n is the frequency-dependent material refractive index, and c is the speed of the light in a vacuum, $dk/d\omega$ can be expressed as follows:

$$\frac{dk}{d\omega} = \frac{n}{c} + \frac{\omega}{c}\frac{dn}{d\omega} = \frac{n + \omega(dn/d\omega)}{c} \tag{1.42}$$

Therefore, the group velocity can be obtained as [5]:

$$v_g = \frac{d\omega}{dk} = \left[\frac{dk}{d\omega}\right]^{-1} = \frac{c}{n - \lambda(dn/d\lambda)} \tag{1.43}$$

This can be rewritten in terms of the group index of a medium N_g as follows:

$$v_g = \frac{c}{N_g} \tag{1.44}$$

Here,

$$N_g = n - \lambda\frac{dn}{d\lambda} \tag{1.45}$$

The refractive index of the dispersive medium and hence the group index are wavelength dependent. Therefore, the phase velocity and group velocity are wavelength dependent.

1.7 Modes in Planar Optical Waveguide

In this section, we will discuss the light behavior inside the planar waveguide, shown previously in Figure 1.1a in order to estimate the propagated modes. We assume that the refractive index of the sandwiched film n_f is greater than those of the upper cover n_c and lower substrate n_s, and also that the refractive index of the substrate is greater than that of the cover. Now, we can define the critical angles for the cover–film interface θ_{1c} and for the film–substrate interface θ_{2c} as follows:

$$\theta_{1c} = \sin^{-1}\left(\frac{n_c}{n_f}\right). \tag{1.46}$$

$$\theta_{2c} = \sin^{-1}\left(\frac{n_s}{n_f}\right). \tag{1.47}$$

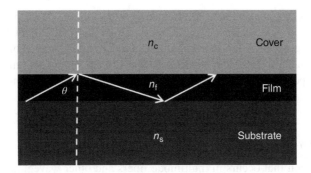

Figure 1.8 Zigzag trajectory of a confinement ray inside the film of a planar waveguide

Based on the assumption that $n_f > n_s > n_c$, we can say that $\theta_{2c} > \theta_{1c}$. Further, based on values of the propagating angle θ of the light inside the film shown in Figure 1.8, we can classify two types of modes: radiation and confinement.

1.7.1 Radiation Modes

These can be obtained in the following two cases:

1. If the propagating angle is less than the critical angle for the cover–film interface ($\theta < \theta_{1c}$), thus the propagating angle will also be less than the critical angle for the film–substrate interface ($\theta < \theta_{2c}$), and hence the radiation will travel in the three zones generating radiation modes.
2. If the propagating angle is greater than the critical angle for the cover–film interface ($\theta > \theta_{1c}$), and less than the critical angle for the film–substrate interface ($\theta < \theta_{2c}$), then, the light cannot penetrate the cover but can penetrate the substrate zone generating also substrate radiation modes.

1.7.2 Confinement Modes

These types of modes can be obtained if the propagating angle is greater than the critical angle for the film–substrate interface ($\theta > \theta_{2c}$), and less than $\pi/2$. In this way, the light cannot penetrate either the substrate or the cover and is totally confined in the film zone generating confinement modes where the light propagates inside the film along a zigzag path.

1.8 Dispersion in Planar Waveguide

The slab waveguide can support number of modes at which each mode can propagate with a different propagation constant. This will occur if the slab waveguide is illuminated by mono-chromatic radiation. It was thought that the axial ray will arrive more quickly than higher-mode rays with longer zigzag paths. However, the group velocity v_g, at which the energy or information is transported, should be taken into account [6]. Additionally, the higher-order modes penetrate more into the cladding, where the refractive index is smaller and the waves travel faster.

The group velocity v_g depends on the frequency ω and the mode propagation constant β. Therefore, the group velocity of a given mode is a function of the light frequency and the waveguide optical properties. It is worth noting that the group velocity v_g of the guided modes is frequency dependent even if the refractive indices of the composing materials are nearly constant. The dependence of the propagation constant and hence group velocity on the frequency can be called **dispersion**.

1.8.1   Intermodal Dispersion

The signal distortion that occurs in multimode fibers and other waveguides is called modal or intermodal dispersion [6]. In this case, the signal is spread in time due to the different propagation velocities for all modes of the optical signal. As a result, a number of allowed propagated modes will be excited through the waveguide when a short-duration light pulse signal is incident on the waveguide. Each mode has its own group velocity. Consequently, a broadened signal will be obtained at the receiver due to the combination of the different modes. To avoid the modal dispersion, single-mode waveguide that allows only one propagated mode can be used.

1.8.2   Intramodal Dispersion

It is worth noting that there is no ideal monochromatic light wave. Therefore, the single mode operation will also consist of various frequencies that constitute the finite spectrum of the source excitation. Each frequency propagates with specific velocity and then arrives the receiver at a given time which results in **waveguide dispersion**. Figure 1.9 shows two different wavelengths (λ_1 and λ_2) that propagate in the core region where ($\lambda_1 < \lambda_2$). As the wavelength increases, the confinement of the mode through the core region will be decreased. Therefore, the penetration of the mode through the cladding region will be increased with higher phase velocity as shown in Figure 1.9.

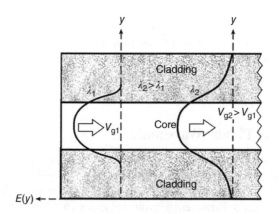

Figure 1.9   Electric field of a fundamental mode through a slab waveguide at different wavelengths (λ_1 and λ_2) where ($\lambda_1 < \lambda_2$)

The refractive index of the waveguide material is wavelength dependent which affects the propagation constant and hence the group velocity. As a result, the propagating light pulse will be broadened which is called **material dispersion** [6]. The combined effect of the waveguide and material dispersion is called **intramodal dispersion**.

1.9   Summary

In this chapter, the definition and different types of optical waveguides are first introduced. Then, Maxwell's equations and the derivation of the wave equation and its solution are presented. Additionally, the basic boundary conditions between two different mediums are discussed. Finally, the phase and group velocity, and types of propagated modes are introduced. Moreover, the dispersion in the planar waveguide is discussed.

References

[1] Sadiku, M.N.O. (2000) *Numerical Techniques in Electromagnetics* (2nd ed.), CRC Press, Boca Raton.

[2] Lifante, G. (2003) *Integerated Photonics Fundemtals*, John Wiely & Sons, Inc., Hoboken.

[3] Joannopoulos, J.D., Meade, R.D. and Winn, J.N. (2001) *Photonic Crystals: Molding the Flow of Light*, Princeton University Press, Princeton.

[4] Ghatak, A. and Thyagarajan, K. (1998) *Introduction to Fiber Optics*, Cambridge University Press, Cambridge.

[5] Pollock, C.R. (ed) (1995) *Fundamentals of Optoelectronics*, Tom Casson, Chicago.

[6] Kasap, S.O. (2012) *Optoelectronics and Photonics: Principles and Practices* (2nd ed.), Prentice Hall, Upper Saddle River.

2

Fundamentals of Photonic Crystals

2.1   Introduction

In the past six decades, manipulating the electrical properties of semiconductor materials has played a huge role in initiating the transistor revolution in electronics, thereby resulting in true a breakthrough in the technology of stimulating highly integrated electronic circuits with high-speed performance. However, the miniaturization of high-speed electronic circuits increases the power dissipation as well as the sensitivity to signal synchronization [1, 2]. In the past few decades, research has begun to turn toward manipulating the optical properties of the materials and utilizing photons instead of electrons as the primary information carrier. Photons can propagate in a dielectric material at much greater speed than electrons in a metallic wire. Further, the bandwidth of the dielectric materials, typically of the order of terahertz, is larger than that of metals, only a few hundreds of kilohertz. Moreover, one of the greatest advantages is that photons are not as strongly interacting as electrons, which reduces the power dissipation in the optical communication systems. Yablonovitch [3] and John [4] were the first to suggest the idea of designing photonic crystal (PhC) materials that manipulate light analogously to the way that ordinary semiconductor crystals manipulate electrons. An electron propagating through the semiconductor crystals sees a periodic potential created by the atomic lattice. In response to the Bragg-like diffraction [1] from the atoms of the semiconductor crystal lattice, electrons with certain energies cannot totally propagate in the semiconductor crystals, thereby generating a gap in the allowed electron energies. The optical analog is PhCs that are composed of dielectric materials with periodic spatial variation of dielectric constants. If the dielectric constants, geometry, and the periodicity of the PhCs are appropriately tuned and if the absorption of light by the dielectric material is minimal, then a range of certain forbidden wavelengths, called a photonic bandgap (PBG), is created. Electromagnetic modes with wavelengths lying within the PBG cannot propagate through the PhCs due to the

Computational Liquid Crystal Photonics: Fundamentals, Modelling and Applications, First Edition.
Salah Obayya, Mohamed Farhat O. Hameed and Nihal F.F. Areed.
© 2016 John Wiley & Sons, Ltd. Published 2016 by John Wiley & Sons, Ltd.

reflection and refraction at the boundaries of the alternating materials. Periodic dielectric structures with bandgaps have many comprehensive applications in different branches of wave optics, such as filters, mirrors, resonators, sensors, and lasers and microwaves.

The scope of this chapter is to provide detailed descriptions of the propagation of light in PhCs. Following this introduction, common configurations for PhC structures are discussed in detail in Section 2.2. Section 2.3 gives a comprehensive description of PBG calculations; the defects in PhCs are presented in Section 2.4; the ordinary and developed fabrication techniques used for fabricating PhCs are introduced in Section 2.5; the applications of PhCs are presented in Section 2.6; and, finally, the description, construction, modes of operation, and fabrication techniques of photonic crystal fibers (PCFs) are given in Section 2.7.

2.2 Types of PhCs

As shown in Figure 2.1, PhCs can be engineered in one, two, or three dimensions to give a PBG for a specific range of frequencies for which electromagnetic waves are forbidden to exist with the crystal [5]. One-dimensional (1D) thin film stacks have been used as optical gratings that selectively reflect based on frequency. Two-dimensional (2D) PhCs have been applied to commercially improve photonic device efficiency with high confinement and low loss; examples include PCFs and PhC vertical cavity surface emitting lasers. Three-dimensional (3D) PhCs attract considerable research interest because the light can be modulated in all directions. It is very important to investigate and understand the PBG structures, also known as dispersion curves for electromagnetic waves propagating in various 1D, 2D, and 3D PhCs.

2.2.1   1D PhCs

1D PhC structures have multiple applications in many branches of physics, such as in creating lower-loss optical fibers and nonlinear optical devices. One of the most common wave optics applications is in interference coatings that create a PBG in the electromagnetic dispersion curves for those electromagnetic waves that propagate perpendicular to the structure. The interference coatings are layered dielectric structures that have a periodicity of a quarter wavelength in their index of refraction. Figure 2.2 shows the simplest form of 1D PhC that consists of two loss-less dielectrics with thickness and refractive indices of a, b and n_1, n_2, respectively.

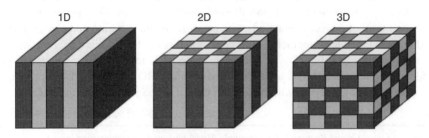

Figure 2.1   Schematics of PhCs. Different shades represent different dielectric materials. *Source*: Ref. [5]

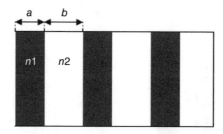

Figure 2.2   Schematic for the 1D PhC structure

These two dielectrics are arranged alternately in 1D. When illuminating the 1D structure from the left with a light beam with wavelength λ, the beam will be reflected at the interfaces between the cells, which results in a reflected light beam. The difference in the optical path d between the incident and reflected light is expressed as follows [6, 7]:

$$d = 2\left(n_1 \times a + n_2 \times b\right). \tag{2.1}$$

The incident light with wavelength λ is totally reflected, namely Bragg reflection, the forbidden transmission through the crystal if the optical path difference satisfies the following condition:

$$d = m\lambda \quad m = 1,2,3,\ldots \tag{2.2}$$

The central wavelength of the bandgap λ_c is expressed as follows [6]:

$$\lambda_c = \frac{2\left(a \times n_1 + b \times n_2\right)}{m} \quad m = 1,2,3,\ldots \tag{2.3}$$

2.2.2   2D PhCs

After the identification of 1D crystals, it took more than one century to add higher dimensional periodic PhCs. 2D systems exhibit most of the important characteristics of PhC structures and are the topic of interest for holey fiber (HF). This type has a periodic geometry in two directions and is homogeneous in the third. Figure 2.3a displays a 2D square lattice of high dielectric constant in a background of low dielectric constant. However, Figure 2.3b shows the square lattice of columns of low dielectric constant in a background of high dielectric constant [1, 2]. Other types of 2D systems with, triangular, honeycomb, and veins lattices are possible, as shown in Figures 2.4 and 2.5a and b, respectively. It is evident From Figure 2.4 that the triangular lattice structure consists of a periodic array of parallel dielectric columns of circular cross section whose intersection with a perpendicular plane form the triangular lattice. The square lattice and the triangular lattice PhCs are of great importance and are now being applied experimentally for various advanced photonic devices [8, 9].

The clef to understand the 2D periodic photonic crystals can be recognized by splitting the fields into two polarization schemes: transverse electric (TE), in which the electric field is perpendicular to the axis of the 2D rods and transverse magnetic (TM), in which the magnetic field is perpendicular to the same axis.

(a)   (b)

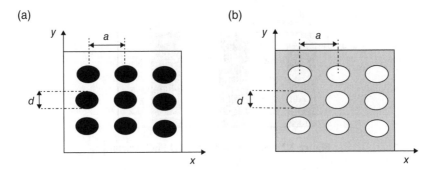

Figure 2.3 Top views for a 2D square lattice PhC structure: (a) high dielectric constant in a background of low dielectric constant and (b) columns of low dielectric constant in a background of high dielectric constant

(a)   (b)

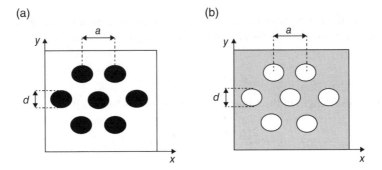

Figure 2.4 Top views for a 2D triangular lattice PhC structure: (a) high dielectric constant in a background of low dielectric constant and (b) columns of low dielectric constant in a background of high dielectric constant. *Source*: Ref. [8]

(a)   (b)

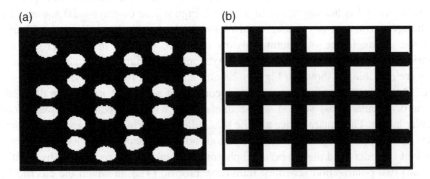

Figure 2.5 Top views of 2D PhC structures having (a) a honeycomb lattice and (b) a veins lattice. *Source*: Refs. [1, 9]

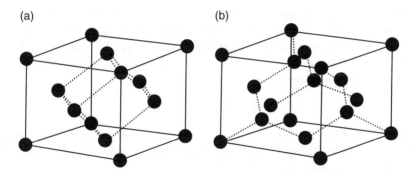

(a)   (b)

Figure 2.6 Schematics for 3D PhCs: (a) fcc lattice and (b) diamond lattice. *Source*: Ref. [2]

Figure 2.7 Inverse woodpile crystal 3D PhCs. *Source*: Refs. [3, 8]

2.2.3 3D PhCs

The 3D PhC has been suggested as a means of controlling electromagnetic waves in 3Ds. This crystal is a dielectric structure that is periodic along all three orthogonal axes. There are many common dielectric topologies that have been identified to uphold complete 3D gaps [2, 10–12]. However, in many applications, a full PBG is not required since, a priori restrictions of the propagation direction and/or polarization may be given, or other mechanisms may be used to suppress propagation in certain directions. Figure 2.6a and b shows two classes of 3D PhCs: fcc lattice and diamond lattice, respectively.

The past decade has been marked by many types of 3D structures such as inverse woodpile and woodpile or layer-by-layer structure. Figure 2.7 shows the inverse woodpile 3D structure that consists of two sets of pores perpendicular to each other.

2.3 Photonic Band Calculations

In this section, we study the propagation of electromagnetic waves in PhCs extending infinitely along all directions of the space. A PhC is characterized by a dielectric constant $\varepsilon_r(x, y, z)$ assumed to be real and periodic along N directions (with $N = 1$, 2, or 3) and invariant along the other $N - 3$ orthogonal directions. The considered crystals are made of nonmagnetic materials.

PhCs consist of regularly repeating internal regions of high and low dielectric constant. Like solid-state crystals, PhCs are categorized by their underlying lattice. A lattice is made of elementary cells that are repeated periodically in space. It can be generated by three basis vectors, $\vec{a}_1$, $\vec{a}_2$, and $\vec{a}_3$ and a set of integers k, l, and m so that each lattice point, identified by a vector $\vec{r}$, can be obtained from

$$\vec{r} = k\vec{a}_1 + l\vec{a}_2 + m\vec{a}_3. \tag{2.4}$$

There are five distinct 2D lattices that when repeated can fill the whole space. Figure 2.8 shows these lattices. Further, there are 14 3D distinct lattices. These distinct types are called Bravais lattices, named after Auguste Bravais [11].

In the next sections, we shall outline the plane wave method [12] for the numerical modeling of propagation in PhCs. The prerequisite of this method is indeed the assumption that the crystal is infinite along all directions of space. This method plays a vital role in the calculation of PBGs [12]. Two other finite-difference time-domain (FDTD) methods that have been developed for calculating the PBG of 2D PhCs will also be presented.

2.3.1 Maxwell's Equations and the PhC

In 1928, the investigations of wave propagation in periodic PhCs were first studied by Felix Bloch [12, 13]. He showed that the propagation of un-scattered waves in such media is governed by a periodic envelope function multiplied by a plane wave. The proposed methodologies by Bloch in quantum mechanics can be used in electromagnetism by formulating Maxwell's equations as an eigenvalue problem in analogy with Schrödinger's equation. Assuming that the charge and the electric current densities are absent, merging the source-free Faraday's and Ampere's laws and reformulating by eliminating E or H, the following time-dependent Maxwell's curl equations for linear, isotropic, nonmagnetic, and nondispersive material can be written as follows:

$$\vec{\nabla} \times \left\{ \frac{1}{\varepsilon(r)} \vec{\nabla} \times \vec{H}(r,t) \right\} = -\frac{1}{c^2} \frac{\partial^2}{\partial t^2} \vec{H}(r,t). \tag{2.5}$$

Assume that, each field has a harmonic time dependence $e^{-i\omega t}$, where ω is the angular frequency; the following eigenvalue equation is obtained for $\vec{H}$:

$$\vec{\nabla} \times \frac{1}{\varepsilon} \vec{\nabla} \times \vec{H} = \left(\frac{\omega}{c} \right)^2 \vec{H}, \tag{2.6}$$

where ε is the dielectric permittivity function $\varepsilon(x, y, z)$ and c is the light speed. The eigenvalue is the wavenumber $k = (\omega/c)^2$ and the eigen operator is

$$O_{\vec{H}} = \vec{\nabla} \times \frac{1}{\varepsilon} \vec{\nabla} \times. \tag{2.7}$$

The two curls equal somewhere to the "kinetic energy" and $1/\varepsilon$ to the "potential" (compared with the Schrödinger Hamiltonian).

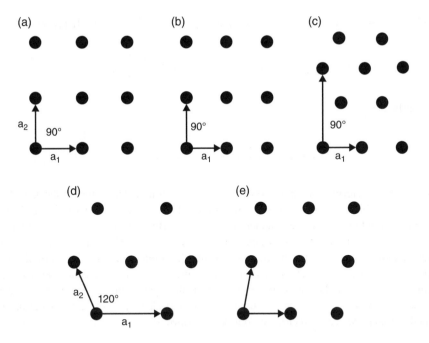

Figure 2.8 The five Bravais lattices of 2D crystals: (a) square, (b) rectangular, (c) centered rectangular, (d) hexagonal, and (e) oblique

As reported by Joannopoulos [13], Eq. (2.6) has some salient features that hold the for general condition, requiring only a real dielectric function [13]. The first feature is that the operator O is a linear operator, where H_1 is a mode and H_2 is also a mode, then $H_1 + H_2$ is also a mode. Further, the operator O is a Hermitian operator, that is (doing the same to the left and right), under the inner product between two fields $(f, Og) = (Of, g)$. In addition, the operator O must have real eigenvalues only if the dielectric function is positive everywhere. Moreover, any two different modes will be orthogonal, that is, $(H_i, H_j) = \delta_{ij}$. Note that, degenerate modes are generally caused by system symmetry and are not necessary orthogonal.

There is a significant difference compared to quantum mechanics in which the constraints $\vec{\nabla} \cdot \vec{H} \neq 0$ and $\vec{\nabla} \cdot \varepsilon \vec{E} \neq 0$ eigen solutions, valid at $\omega = 0$, are excluded. The solutions of static field solutions without magnetic or electric charges are forbidden [13].

2.3.2 Floquet–Bloch Theorem, Reciprocal Lattice, and Brillouin Zones

The periodic dielectric function of the PhC structure is given by

$$\varepsilon(\vec{x}) = \varepsilon(\vec{x} + \vec{R}_i), \tag{2.8}$$

where, $\vec{R}$ is the primitive lattice vector and i terms for the periodicity in 3Ds and equals to 1, 2, and 3. According to Bloch theorem, the solution to the eigenvalue equation (2.6) can be given as follows:

$$\vec{H}(\vec{x}) = \exp\left(j\vec{k} \cdot \vec{x}\right) \vec{H}_{n,k}(\vec{x}), \tag{2.9}$$

with eigenvalues $\omega_n(\vec{k})$, where $\vec{H}_{n,k}(\vec{x})$ is a periodic envelope function satisfying:

$$\left(\vec{\nabla} + j\vec{k}\right) \times \frac{1}{\varepsilon}\left(\vec{\nabla} + j\vec{k}\right) \times \vec{H}_{n,k} = \left(\frac{\omega_n(\vec{k})}{c}\right)^2 \vec{H}_{n,k}. \tag{2.10a}$$

Similarly, for $\vec{E}$

$$\left(\vec{\nabla} + j\vec{k}\right) \times \left(\vec{\nabla} + j\vec{k}\right) \times \vec{E}_{n,k} = \left(\frac{\omega_n(\vec{k})}{c}\right)^2 \varepsilon(r)\vec{E}_{n,k}, \tag{2.10b}$$

yielding different Hermitian eigen problems over the primitive cell of the lattice at each Bloch wavevector $\vec{k}$. This primitive cell is a limited area if the frame is periodic in all directions, leading to separated eigenvalues labeled as $n = 1, 2\dots$. Band structure or dispersion diagram can be obtained by plotting these eigenvalues $\omega_n(\vec{k})$ versus $\vec{k}$. This diagram is composed of discrete bands that map out all periodic available interactions in the periodic system. Additionally, $\vec{k}$ is not necessarily real; complex $\vec{k}$ values result in evanescent modes that cannot remain in the crystal and decay exponentially near the crystal boundaries. Also, according to the Bloch theorem, the eigen solutions also vary periodically with $\vec{k}$: the solution at $\vec{k}$ is repeated at $\vec{k} + \vec{G}_j$, where $\vec{G}_j$ vectors are defined as follows:

$$\vec{R}_i \cdot \vec{G}_j = 2\pi\ \delta_{ij}. \tag{2.11}$$

The vectors $\vec{G}_j$ are called the primitive reciprocal lattice vectors. It can be easily shown that for a 3D lattice,

$$\vec{G}_1 = 2\pi\ \frac{\left(\vec{R}_2 \times \vec{R}_3\right)}{\vec{R}_1 \cdot \left(\vec{R}_2 \times \vec{R}_3\right)}. \tag{2.12a}$$

$$\vec{G}_2 = 2\pi\ \frac{\left(\vec{R}_3 \times \vec{R}_1\right)}{\vec{R}_2 \cdot \left(\vec{R}_3 \times \vec{R}_1\right)}. \tag{2.12b}$$

$$\vec{G}_3 = 2\pi\ \frac{\left(\vec{R}_1 \times \vec{R}_2\right)}{\vec{R}_3 \cdot \left(\vec{R}_1 \times \vec{R}_2\right)} \tag{2.12c}$$

As an example, for the case of a 2D crystal with a period a along the x-axis and with the self-same period a along an axis rotated from 60° with respect to the x-axis as shown in Figure 2.8. The two vectors $\vec{R}_1$ and $\vec{R}_2$ that can be used to form the direct lattice are defined as follows:

$$\vec{R}a_1 = \hat{x}, \tag{2.13a}$$

$$\vec{R}a_2 = \left(\frac{\hat{x}}{2} + \frac{\hat{y}\sqrt{3}}{2}\right). \tag{2.13b}$$

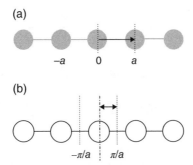

(a)

$-a$ 0 a

(b)

$-\pi/a$ π/a

Figure 2.9 1D photonic crystals: (a) the network of lattice points in real space. (b) The Brillouin zone and irreducible Brillouin zone are defined by the distances $[-\pi/a, +\pi/a]$ and $[0, +\pi/a]$, respectively

Further, the two vectors $\vec{G}_1$ and $\vec{G}_2$ that can be used to form the reciprocal lattice in the Fourier space or $\vec{k}$-space are defined as follows:

$$\vec{G}_1 = \frac{4\pi}{a\sqrt{3}}\left(\frac{\hat{x}\sqrt{3}}{2} - \frac{\hat{y}}{2}\right),$$ (2.14a)

$$\vec{G}_2 = \frac{4\pi}{a\sqrt{3}}\,\hat{y}.$$ (2.14b)

The Fourier expansion of the $\vec{H}$ field given by Eq. (2.9) is defined by

$$\vec{H}(x,y) = \exp\left(j\vec{k}\cdot\vec{r}\right)\sum_{m=-\infty}^{+\infty}\sum_{n=-\infty}^{+\infty}\vec{H}_{m,n}\exp\left\{j\left(m\vec{G}_1 + n\vec{G}_2\right)\cdot\vec{r}\right\}.$$ (2.15)

In response to the periodicity in the $\vec{k}$-space, only the eigen solutions for $\vec{k}$ within the primitive cell of the reciprocal lattice, called the first Brillouin zone, should be computed. For example, in 1D periodic system, where $\vec{R}_1 = a$ for some periodicity a and $\vec{G}_1 = 2\pi/a$, the first Brillouin zone is defined by the region $\vec{k} = -\pi/a$ to $\vec{k} = \pi/a$; all other wavevectors are equal to some points in the zone with the aid of translation by a multiple of $\vec{G}_1$. Further, the first Brillouin zone may itself be excessive if the crystal has additional similarities such as mirror planes; by removing these redundant zones, the *irreducible* Brillouin *zone* is obtained. Thus, in the considered 1D example shown in Figure 2.9, the irreducible Brillouin zone would be $\vec{k} = 0$ to $\vec{k} = \pi/a$.

Similarly, Figures 2.10 and 2.11 show some examples of 2D PhCs with their corresponding Brillouin zones. It can be seen from the figures that one can construct the Brillouin zone and the irreducible Brillouin zone by drawing perpendicular bisecting planes of every lattice vector joining the origin to the nearest vertices of the reciprocal lattice. Each bisector divides the reciprocal space into two half spaces, the origin being contained in one of these half spaces. Then the first Brillouin zone is defined as the intersection of all half spaces which contain the origin. The cell formed in this way is known to physicists as the Wigner–Seits (WS) primitive cell [11]. The triangle, represented in Figure 2.10, constitutes the irreducible Brillouin zone of the square lattice, which is the first Brillouin zone reduced by all of the

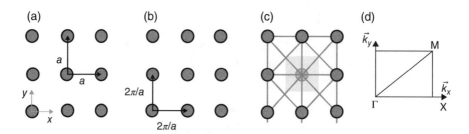

Figure 2.10 2D photonic band crystal (PBC)—square lattice: (a) lattice points in real space; (b) reciprocal lattice; (c) the Brillouin zone, the gray area centered on the origin; and (d) the irreducible Brillouin zone

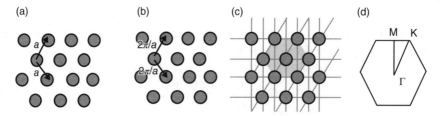

Figure 2.11 2D PBC—triangular lattice: (a) lattice points in real space; (b) reciprocal lattice; (c) the Brillouinzone, the gray area centered around the origin; and (d) the irreducible Brillouin zone

symmetries in the point group of the lattice. Also, it can be demonstrated that the irreducible Brillouin zone of the triangular lattice is the half of an equilateral triangle with apexes Γ, K, and M as shown in Figure 2.11. Here again, the assumption is made that it suffices to let the wavevector $\vec{k}$ describe the edges of the irreducible Brillouin zone along the directions KΓ, ΓM, and MK for reaching the extreme of $\omega(\vec{k})$ which is of interest.

2.3.3 Plane Wave Expansion Method

The plane wave expansion (PWE) method [13–17] is a straightforward way for solving the eigenvalue equations (2.13) and (2.14). The method is based on expanding all the periodic functions into a Fourier series over the reciprocal lattice $\vec{G}$, thereby transforming the differential equations into an infinite matrix eigenvalue problem, which is then truncated and solved numerically. Reference [15] presents MATLAB codes for applying the method to 1D and 2D crystals. In this subsection, a brief description of the application of the PWE for linear, lossless, anisotropic, and periodically inhomogeneous dielectric materials is presented.

The relative permittivity in Maxwell's equations is a periodic satisfying (2.11), namely

$$\overline{\overline{\varepsilon}}\left(\vec{r}\right) = \overline{\overline{\varepsilon}}\left(\vec{r} + \vec{R}_i\right),\tag{2.16}$$

where $\vec{R}$ is Bravais lattice vector.

The electric field $\vec{E}(\vec{r})$ can be expanded in plane (Bloch) waves as follows:

$$\vec{E}\left(\vec{r}\right) = \sum_{\vec{G}}\vec{E}_{\vec{G}}\exp\left(j\left(\vec{k} + \vec{G}\right)\cdot\vec{r}\right),\tag{2.17}$$

where $\vec{G}$ are reciprocal lattice vectors and $\vec{k}$ locates in the irreducible zone of the first Brillouin zone in reciprocal lattice space. $\varepsilon(\vec{r})$ can be expressed in a Fourier series as follows [17]:

$$\bar{\bar{\varepsilon}}(\vec{r}) = \sum_{\vec{G}} \bar{\bar{\varepsilon}}_{\vec{G}} \exp(j\vec{G}\cdot\vec{r}). \tag{2.18a}$$

$$\bar{\bar{\varepsilon}}_{\vec{G}} = \frac{1}{V} \int_{\text{WS cell}} \bar{\bar{\varepsilon}}(\vec{r})\exp(-j\vec{G}\cdot\vec{r}), \tag{2.18b}$$

whose integral region is a WS cell with volume V. Substituting Eqs. (2.17) and (2.18a) into Eq. (2.14), the eigenvalue equation for the electric field $\vec{E}$ has the form [17]

$$\left(\vec{k}+\vec{G}\right)\times\left(\vec{k}+\vec{G}\right)\times\vec{E}_{\vec{G}+\vec{k}} + \frac{\omega^2}{c^2}\sum_{\vec{G}}\bar{\bar{\varepsilon}}_{\vec{G}-\vec{G}'}\cdot\vec{E}_{\vec{G}'+\vec{k}} = 0, \quad \text{for each } \vec{G}. \tag{2.19}$$

Similarly, by using the Fourier expansion for $\bar{\bar{\eta}}(\vec{r}) = \bar{\bar{\varepsilon}}^{-1}(\vec{r})$

$$\bar{\bar{\eta}}(\vec{r}) = \sum_{\vec{G}} \bar{\bar{\eta}}_{\vec{G}} \exp\left(j\vec{G}\cdot\vec{r}\right), \tag{2.20a}$$

with its expansion coefficients:

$$\bar{\bar{\eta}}_{\vec{G}} = \frac{1}{V} \int_{\text{WS cell}} \bar{\bar{\eta}}(\vec{r}) \exp\left(-j\,\vec{G}\cdot\vec{r}\right). \tag{2.20b}$$

The eigenvalue equation in reciprocal lattice space for the vector field $\vec{H}(\vec{r})$ reduces to [17]

$$\left(\vec{k}+\vec{G}\right)\times\sum_{\vec{G}'}\bar{\bar{\eta}}_{\vec{G}-\vec{G}'}\left[\left(\vec{k}+\vec{G}\right)\times\vec{H}_{\vec{k}+\vec{G}'}\right] + \frac{\omega^2}{c^2}\vec{H}_{\vec{k}+\vec{G}} = 0, \quad \text{for each } \vec{G}. \tag{2.21}$$

The eigenvalue equation (2.29) of $\vec{H}$ can be rewritten as given below [17]:

$$\sum_{\vec{G}}\bar{\bar{Y}}_{\vec{G}-\vec{G}'}\cdot\vec{H}_{\vec{k}+\vec{G}'} = -\frac{\omega^2}{c^2}\vec{H}_{k+\vec{G}} \quad \text{for each } \vec{G} \tag{2.22}$$

where

$$\bar{\bar{Y}}_{\vec{G}-\vec{G}'} = \left(\vec{k}+\vec{G}\right)\times\bar{\bar{\eta}}_{\vec{G}-\vec{G}'}\times\left(\vec{k}+\vec{G}'\right). \tag{2.23}$$

Denoting

$$Z = \left[\vec{H}_{\vec{k}+\vec{G}_1} \quad \vec{H}_{\vec{k}+\vec{G}_2} \quad \vec{H}_{\vec{k}+\vec{G}_3} \quad \vec{H}_{\vec{k}+\vec{G}_N}\right]^{\text{T}}, \tag{2.24a}$$

$$Y = \begin{bmatrix} \bar{\bar{Y}}_{\vec{G}_1-\vec{G}_1} & \bar{\bar{Y}}_{\vec{G}_1-\vec{G}_2} & \cdots & \bar{\bar{Y}}_{\vec{G}_1-\vec{G}_N} \\ \vdots & \bar{\bar{Y}}_{\vec{G}_2-\vec{G}_2} & \cdots & \bar{\bar{Y}}_{\vec{G}_2-\vec{G}_N} \\ & & \cdots & \\ \bar{\bar{Y}}_{\vec{G}_N-\vec{G}_1} & \bar{\bar{Y}}_{\vec{G}_N-\vec{G}_2} & \cdots & \bar{\bar{Y}}_{\vec{G}_N-\vec{G}_N} \end{bmatrix}. \tag{2.24b}$$

Then Eq. (2.22) has the following form:

$$Y \cdot Z = -\frac{\omega^2}{c^2} Z. \tag{2.25}$$

This is the matrix form of the H eigenvalue equation, that is, an eigenvalue matrix equation. By solving the eigenvalue equation (2.25), the dispersive relation $k-\omega$ may be abstracted. In addition, the integrals of the Fourier coefficients $\bar{\bar{\varepsilon}}_G$ of $\bar{\bar{\varepsilon}}(\bar{r})$ and $\bar{\bar{\eta}}_G$ of $\bar{\bar{\eta}}(\bar{r})$ must be estimated. Let $\bar{\bar{\xi}}(\bar{r})$ represent $\bar{\bar{\varepsilon}}(\bar{r})$ or $\bar{\bar{\eta}}(\bar{r})$, and using subscripts "a" and "b" represent the inclusion and the host materials of the PBG structure, respectively, then the dyadic $\bar{\bar{\xi}}(\bar{r})$ can be expressed for any field point and each lattice vector $\bar{R}$ as [17]

$$\bar{\bar{\xi}}(\bar{r}) = \bar{\bar{\xi}}_b + \left(\bar{\bar{\xi}}_a - \bar{\bar{\xi}}_b \right) \sum_{\bar{R}} u(\bar{r} - \bar{R}) \tag{2.26}$$

where the 3D step function

$$u(\bar{r} - \bar{R}) = \begin{cases} 0, & \text{when } (\bar{r} - \bar{R}) \text{ lies in the host material} \\ 1, & \text{when } (\bar{r} - \bar{R}) \text{ lies in the inclusion material} \end{cases} \tag{2.27}$$

Hence, the expression (2.18) for $\bar{\bar{\varepsilon}}_G$ becomes:

$$\begin{aligned} \bar{\bar{\xi}}_G &= \frac{1}{V} \int_{\text{WScell}} \bar{\bar{\xi}}(\bar{r}) e^{-j\bar{G}\cdot\bar{r}} dV \\ &= \frac{1}{V} \int_{\text{WScell}} \left[\bar{\bar{\xi}}_b + \left(\bar{\bar{\xi}}_a - \bar{\bar{\xi}}_b \right) \sum_{\bar{R}} u(\bar{r} - \bar{R}) \right] e^{-j\bar{G}\cdot\bar{r}} dV \\ &= \begin{cases} \bar{\bar{\xi}}_b + \Omega\left(\xi_a - \xi_b \right) & \text{when } \vec{G} = \vec{0} \\ \frac{1}{V}\left(\xi_a - \xi_b \right) \int_{\text{inclusion}} e^{-j\bar{G}\cdot\bar{r}} dV, & \text{when } \vec{G} \neq \vec{0} \end{cases} \end{aligned} \tag{2.28}$$

where $\Omega = V_a/V$ is defined as the filling fraction, V_a is the volume of inclusion in a WS cell, and the integral on the WS cell is zero for $\bar{\bar{G}} \neq \bar{\bar{0}}$. In the case of an isotropic material as (where $i = a$ or b) for the E-equation; $\bar{\bar{\varepsilon}}_i = \varepsilon_i \bar{\bar{I}}$ or $\bar{\bar{\eta}} = \varepsilon_i^{-1} \bar{\bar{I}}$ (where $i = a$ or b) for the H equation, $\bar{\bar{\xi}}$ can be degenerated appropriately and the characteristics of the PBG structure with respect to dyadic permittivity will be associated with the scalar dielectric constants ε_a and ε_b.

Generally, the plane wave may be decomposed into TM and TE waves with respect to a specific vector. In the case of a 2D rod structure, the TM and TE waves are with respect to the axis of the rod. When Eq. (2.28) is employed, the matrix size will be $3N \times 3N$ due to three components of E in the 3D case, or $2N \times 2N$ due to two components of E in the 2D case with TE wave propagation, or $N \times N$ due to one component of E in the 2D case with TM wave propagation. To reduce the computing time, Eq. (2.25) is employed, where the matrix size will be $N \times N$ due to one component of H in the 2D case with TE wave. The benefit from employing both E and H equations is that the fields for TM and TE waves can be calculated independently from the matrix equation with same the size of $N \times N$.

2.3.4 FDTD Method

Employing the PWE method frequently affects the numerical stabilities because of the discontinuity of the permittivity of the crystal as well as certain components of the electric field [10]. Strictly speaking, the truncation of the Fourier series of discontinuous functions results in a tardy convergence of their development whatever be the order of the truncation. There are many numerical methods that have been developed for investigating the PBG in the 2D PhC. However, these methods are more complex than the plane wave method; they generally provide high numerical stabilization. The best known of these methodologies are referred to as finite FDTD. In the following subsections, two different approaches based on FDTD can be applied to investigate the PBG of 2D PhCs: band structure and transmission diagram.

2.3.4.1 Band Structure

Bloch periodic boundary conditions are used to bound the FDTD computational window to calculate the properties of an infinite periodic structure by simulating only one unit cell. The photonic band structure is calculated by determining the angular frequencies, $\omega_n(\vec{k})$, as a function of the wavevector $\vec{k}$ [13]. Figure 2.12 shows a flow chart that describes the steps for calculating the band structure.

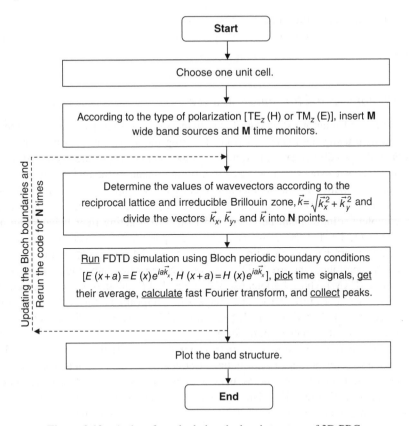

Figure 2.12 A chart for calculating the band structure of 2D PBCs

2.3.4.2 Transmission Diagram

In this method, a Gaussian pulse will be assumed to propagate normally (normal incidence) to a 2D PhC structure that will be bounded by nonphysical walls, a perfectly matched layer (PML). Then, the propagation characteristics around the crystal are investigated. The transmission coefficient is computed by comparing the incident and the transmitted fields as given by the following relation [10]:

$$T(\omega_1) = \frac{|E(y_{\text{trans}}, \omega_1)|}{|E(y_{\text{inc}}, \omega_1)|}. \tag{2.29}$$

Here, y_{inc} is a value of y at the very beginning of the domain and y_{trans} at the very end.

Some numerical problems may arise whenever computing transmission using Eq. (2.29), due to, for example, unexpected reflections. An alternative way is to use a reference structure, and hence the transmission T becomes thus:

$$T(\omega_1) = \frac{|E(y_{\text{trans}}, \omega_1)|}{|E^{\text{ref}}(y_{\text{trans}}, \omega_1)|}. \tag{2.30}$$

This is the fraction of the field that passes through an area at $y = y_{\text{trans}}$ with and without a specific subject. The transmission diagram (T versus ω) indicates the bandgaps if they exist.

2.3.5 Photonic Band for Square Lattice

The photonic band diagram showing all the dispersion characteristics $\omega(\vec{k})$ of the PhC is investigated here using the FDTD method. By using the same concepts used in solid-state physics [15], the possibility exists to draw upon the symmetries of the crystal to restrict the calculation to the wavevectors k contained in a finite domain of the reciprocal space donated as the irreducible Brillouin zone as mentioned earlier. The importance of the Brillouin zone stems from the Bloch wave description of waves in a periodic medium, in which it is found that the complete solutions can be calculated from investigating their behavior in a single Brillouin zone [1, 10, 16]. Here, we consider the square lattice structures that are formed by dielectric rods arranged in air with $d/a = 0.36$ and $n = 3.4$ [16], as shown in Figure 2.13a. As previously mentioned, the irreducible Brillouin zone of this lattice is an isosceles, right-angled triangle (ΓXM) whose hypotenuse ΓM is parallel to the diagonal of each elementary square, as shown in Figure 2.13. The dispersion diagram is formed by pursuing the different directions ΓX, XM, and ΓM, beginning with the vertices of the reciprocal lattice located at increasing distances from the origin.

It is evident from the dispersion diagram shown in Figure 2.13c that there is a complete TM$_z$ bandgap over a significant range of normalized frequencies, which extends from $\lambda = a/0.302$ to $\lambda = a/0.443$. The transmission across the array with width 10×10 of the considered PhC structure is calculated in Figure 2.13d. The figure indicates that there is one large PBG for TM$_z$ polarization that extends from $\lambda = a/0.302$ to $\lambda = a/0.443$. Extensive research has concluded that large, complete TM$_z$ bandgaps are most likely to be found using isolated regions of high

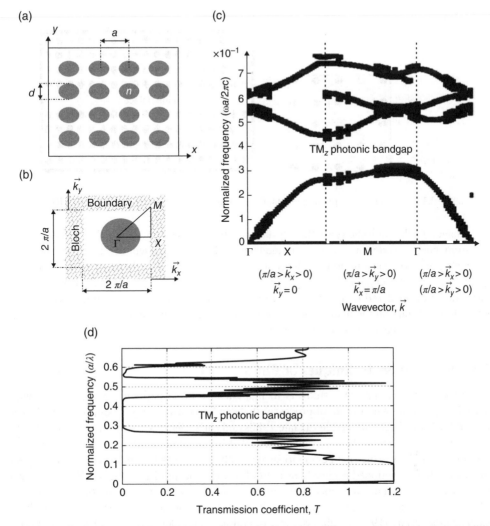

Figure 2.13 Bandgap investigation: (a) structure details, $d=0.36a$, $a=0.58\,\mu m$, and $n=3.4$; (b) a unit cell of $\vec{k}$ space; (c) band structure for M_z mode; and (d) transmission across the photonic bandgap structure

dielectric constant material imbedded in a background of lower dielectric constant, whereas for TE_z bandgaps, significant bands are found for isolated regions of lower dielectric constant immersed in a higher dielectric background [16].

2.4 Defects in PhCs

The implementation of various optical devices based on PhC platforms mostly necessitates the enrollment of disorders that cause a disruption in the periodicity of the crystal. Figure 2.14 illustrates two common kinds of defects: point and extended. Point defects are very local

(a)   (b)

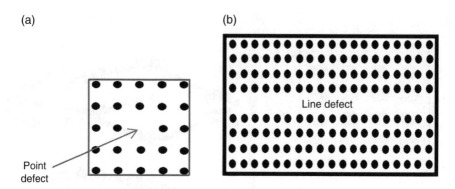

Figure 2.14 Top views for planar defects: (a) point defect and (b) line defect. *Source*: Ref. [9]

disturbance in the periodicity of the crystal, and their presence leads to the appearance of electromagnetic modes at discrete frequencies. Whereas extended defects may result in the appearance of transmission bands in spectral regions and may be seen as analogous to dislocations of the crystal. One of the most intriguing properties of PBG crystals is the appearance of localized defect modes within the PBGs when a defect is introduced into a perfect PhC. Studying the nature and the properties of the localized modes is very important for creating novel nano-photonic devices [9].

2.5   Fabrication Techniques of PhCs

The major defiance for PhCs study is to manufacture these periodic structures [18] with sufficient precision with the aid of processes that can be robustly mass produced. The fabrication of PhCs had been initiated in the late 1980s and early 1990s with few practical results. In large part, this is because there was no cheap, reliable way of fabricating the crystals. In 1987, Yablonovitch [3] was the first person to successfully create a PhC exhibiting a 3D PBG, and this experiment functioned primarily as a proof of the concept of PhC theory. He has chosen his lattice constant in millimeters rather than nanometers to verify his prediction of a complete 3D PBG [3]. Various techniques have been proposed to manufacture different forms of PhCs such as lithography [19, 20], self-assembly of colloidal particles [21, 22], holographic lithography (HL) [23, 24], direct laser writing [25], nano-imprint lithography [26], and so on. The scope of this section is to shed light on the recent methods used for fabricating nano-PhCs and to evaluate their potential usefulness.

2.5.1   Electron-Beam Lithography

As early as 1994, one of the most popular techniques applied for fabricating PhC, known as layer-by-layer lithography [19, 20], was discovered. This method is based on etching a cross section of the PhCs pattern onto a substrate, backfilling the etched holes with SiO_2 and then depositing another layer of substrate. This process is repeated for each desired cross section of the PhCs pattern. After depositing a sufficient number of layers (usually a minimum of 10 in order to exhibit the desired bandgap), the silica is dissolved, leaving a PhC that can have a PBG close to the visible regime. The etching process is accomplished through electron beam (e-beam) lithography (EBL) [19]. In this process, a beam of electrons is used to drill holes in a given substrate. The advantage of the process stems from the simplicity of introducing defects into the design, since each hole must be drilled sequentially. On the other hand, prohibitive costs and slow writing speeds are considered as two major disadvantages to EBL.

2.5.2   Interference Lithography

X-ray interference lithography (IL) [20] is considered as an alternative lithographic technique that overcomes the pitfalls of the EBL technique. IL uses the interference pattern created by several (typically 4–6) high-frequency beams to imprint the desired pattern onto a photosensitive resin. After a particular layer is completed, the resin is exposed to ultraviolet (UV) light that hardens it. Then, the next layer is deposited and the process is repeated. This method overcomes the disadvantages of the EBL technique being about 10% of the cost and also incredibly rapid. Furthermore, due to the use of small wavelengths, the resolution of the etched pattern is superb. In Ref. [20], a multiple-exposure two-beam interference method was successfully applied to fabricate any 1D, 2D, and 3D periodic structures that cannot be fabricated by a three- or four-beam interference mechanism. The 1D structures could be oriented in any direction in space. By merging multiple 1D structures each with appropriate orientation angles, any desired 3D periodic structure can be easily created. Moreover, the desired lattice constants of the created 3D periodic structures can be controlled using any value of the θ-angle between two laser beams, which is difficult to achieve by the commonly used one-exposure multi-beam interference technique [20]. As indicated by Vogelaar [20], the lattice constant of a two-beam interference pattern can be calculated by $\Lambda = \lambda/2\sin\theta$, where λ is the excitation wavelength. In the experiment in Ref. [20], a 50/50 beam splitter was used to get two laser beams of the same profile, equal polarization, and equal intensity. In addition, the two beams are redirected to the sample using two mirrors and the resultant interfered beam is periodically modulated in one direction. The angle between the two laser beams could be easily ruled by rotating the two mirrors. A sample was fixed in a double rotation stage, which could be rotated around the normal z-axis around the horizontal y-axis. There are numerous types of excitation lasers such as He–Cd laser emitting at 325 nm and at 442 nm or an argon laser emitting at 514 nm that can be used based on the used photoresist and structure to be manufactured. These lasers show stabilization and have a long coherence length. According to the excitation laser wavelength, different types of photoresists, such as SU8 and JSR (negative photoresists) and AZ4620 and S1818 (positive photoresists) have been used. There are two main problems with X-ray lithography. The first one arises in the generation of a correct interference pattern due to the difficulty in calculating the appropriate beam parameters. Recently, this limitation has largely been overcome, and several common crystalline patterns have been fabricated [20]. The second one is the difficulty in introducing defects due to the uniformity of the interference pattern.

2.5.3   Nano-Imprint Lithography

Nano-imprint lithography (NIL) [26] has now been used for fabricating complex micro-nanoscale PhC structures with minimum cost, large throughput, and high efficiency. In response to these superior features, it was adopted by the International Technology Roadmap for Semiconductors (ITRS) in 2009 and applied for industrial fabrication in 2013. Further, Toshiba has used NIL for 22 nm and beyond. NIL is anticipated to play an important role in the commercialization of nanostructure devices [26]. There are two main NIL processes: UV-NIL with a rigid template and soft UV-NIL using a soft mold. Soft UV-NIL is adopted due to its cheapness and insensitivity to particle contaminants. In addition, it has low surface energy of adaptable mold materials and low imprinting. Further, UV-NIL utilizes gradually sequential micro-contact and a "peel-off" separation method for thin-film-type molds. In response to these outstanding features, soft UV-NIL can be applied for obtaining large-area patterns on nonplanar surfaces and even bent substrates and for fabricating complex 3D micro-/nanostructures with high-aspect-ratio features. Although, the soft mold is preferred, it also has some potential drawbacks. One of the major drawbacks that limits the resolution of imprinted patterns is the deformation of soft molds due to the comparatively low Young's modulus. Moreover, the poor solvent resistance and the effect of pressure and thermal expansions lead to imprinted patterns with poor dimensional stability as well as soft molds with relatively low mold lifetime. The progress of forming micro-/nanostructures by soft UV-NIL begins with the production of a master template using EBL, IL, or other patterning technologies then replicating the soft mold by this master template. After that, the sample is imprinted in the UV-curable resist using the replicated soft mold, and transferring the replicated patterns from the UV-curable resist to the substrate or functional materials by etching or a lift-off process. The removal of the residual layer is typically performed by using a reactive ion etching (RIE) process with oxygen plasma. These steps jointly affect the quality of the final process in terms of resolution, uniformity, fidelity, patterning area, and line edge roughness.

2.5.4   Colloidal Self-Assembly

Over the past decades, the colloidal self-assembly approach has been considered as a promising and practical approach for the fabrication of periodic and non-periodic photonic nanostructures [25]. This method depends on the interactions between the colloidal particles, and the colloidal structures can be affected dramatically and modulated by applying an additional external field. Compared to other approaches, colloidal self-assembly is quick, efficient, and cheap. Recently, the colloidal self-assembly approach has been applied for producing different nanoscale structures such as displays, optical devices, and photochemical and biological sensors.

2.6   Applications of PhCs

The PhC can manipulate the photons and thus has received great interest in different applications, such as sub-wavelength imaging, scanning photon tunneling microscopy, phase shifters, logic gates, and optical routers. The design of PhC-based applications is based on the bandgap or defect engineering. In this regard, the light can be controlled by introducing a defect in the periodic structure. If a line defect is introduced, the light can be transmitted through such a small waveguide along the defect only [27]. Alternatively, a point defect can be introduced into the PhC, resulting in a photonic nano-cavity [28, 29]. Additionally, line and point defects can be combined to form ultra-small photonic circuits in different applications [30–32].

Moreover, the bandgap of the PhC can be engineered to suppress spontaneous emission to increase the performance of various photonic devices. Further, the band edge of the PhC can be controlled [33, 34], so that the group velocity of light becomes zero. At the band edge, a standing wave is obtained by Bragg reflection of light propagating in various directions. The formulated standing wave looks like a cavity mode that can be used for laser applications. PhCs can also control the propagation of light through the structure. This interesting characteristic can be used to slow the propagation speed, change propagation direction, and induce negative refraction. As a result, novel photonic devices can be realized for imaging, optical switching, routing, and power splitting.

PhCs have high flexibility control over photons. Therefore, they could play an important role in increasing the capacity of optical communication networks. The PhCs could be used to design lasers with very narrow wavelength range on the same chip with highly selective optical filters [28]. It should also be noted that there has been a dramatic improvement in the fabrication process of the PhCs. Therefore, 3D PhCs are expected to play an important role in the forthcoming decade. This is due to the control of the light over three dimensions instead of the current 2D PhCs.

PCFs or holey fibers (HFs) [35] have uncommon characteristics which cannot be achieved by conventional optical fibers. The PCF usually consists of silica with micro structured air holes in the cladding region running along its length. The PCFs can transmit a wide range of wavelengths without suffering from dispersion. In addition, the dispersion can be tailored until zero or flat dispersion is obtained. Further, PCFs can support a single mode over a wide range of wavelengths with large core region. The PCFs can be used for a wide range of applications such as polarization splitters, polarization rotators, filters, and multiplexers–demultiplexers. In the following section, the main characteristics of the PCFs will be introduced.

2.7 Photonic Crystal Fiber

The PhC can be used for waveguide applications due to the capability of designing waveguides that can trap or prohibit the propagation of light at any wavelength including microwaves, infrared waves, and visible light waves. The prohibition of propagation of light along a certain direction in the periodic PhC is due to the destructive interference of reflected and refracted parts of the light in a specific frequency range. The PCF is an example of PhC waveguide with a wide range of applications.

2.7.1 Construction

The PCF [35] is one of the most recent advances in optical fiber technology that has attracted considerable interest from many researchers around the world. It can be fabricated from a single material while conventional optical fibers are manufactured from two or more materials. The PCF, as shown in Figure 2.15a, has a core of pure silica with a microstructure of air holes in the cladding region [36]. Figure 2.15 shows a schematic diagram of a PCF with (a) a solid core or (b) a hollow core. PCFs can be endlessly single mode over a wide wavelength range [37], have a large effective mode area [38], and can be tailored to achieve nearly zero and flat dispersion over a wide range of wavelengths [39].

(a)   (b)

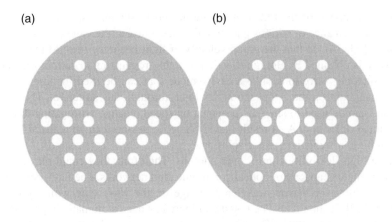

Figure 2.15   Schematic diagram of (a) an index-guiding PCF and (b) photonic bandgap-guiding fiber

2.7.2   Modes of Operation

The light is guided through the core region in the conventional optical fiber by total internal reflection. However, PCFs, as shown in Figure 2.15, can be grouped into two different categories, based on the light-guiding mechanisms in the fibers: high index-guiding PCFs and PBG-guiding fibers.

2.7.2.1   High Index Guiding Fiber

Index-guiding PCFs [36], also known as HFs, have a solid core and can guide the light due to the principle of modified total internal reflection (MTIR). This mechanism is similar to the light guiding mechanism through the conventional step-index fibers. However, the HF cladding index is strongly wavelength dependent due to the wavelength scale features of the HFs. In addition, the air holes along the HF, as shown in Figure 2.15a, decrease the effective refractive index of the cladding region. Therefore, the light will be confined to the solid core with a relatively higher index.

2.7.2.2   PBG Fibers

PBG fibers [40] can guide the light through a low index core (say air), as shown in Figure 2.15b, due to the PBG effect. In this mechanism, the frequencies located in the bandgap are trapped in the low index core and cannot leak through the cladding. This interesting new technology could be used as a basis for high power delivery without nonlinear effects or material damage. In addition, the hollow core can be filled with different materials such as gases, particles, liquid, and liquid crystals that can be used for sensing applications. Further, the mode will be well confined through the air core, and hence the PBG fibers are less sensitive to bending. Additionally, the PBG are radiation insensitive making them suitable for radiation hard environments.

2.7.3 Fabrication of PCF

The fabrication of PCF [41, 42] is similar to that of the conventional fiber fabrication starting with a fiber preform. The modified chemical vapour deposition or vapor axial deposition process is usually used to form the conventional preforms. The PCF preforms are based on stacking a number of capillary silica tubes and rods to form the desired microstructure design. The stack is then fused into a preform and pulled to a fiber at a sufficiently low temperature (~1950°C) to avoid hole collape. The desired preform is then drawn to a fiber in a conventional high-temperature drawing tower to produce the required PCFs in kilometer lengths. This technique can control the PCF structure geometrical parameters as well as the index profile throughout the cladding region. In addition, the air holes arrangement through the drawing process can be controlled, and hence complex PCF designs with high air-filling ratio can be produced. The final fabrication step is the coating of the PCF with a protective standard jacket to allow robust handling of the PCF as well as preventing PCF micro-bending.

2.7.4 Applications of PCF

The interesting features of the PCFs that cannot be provided by the conventional optical fiber offer a number of different applications in a wide area of science and technology. The PCF has endlessly single-mode guiding with a large effective mode area. Therefore, PCFs can be used to transmit high laser powers without structure damage. Additionally, the large effective mode area offers low power density that minimizes the nonlinear optical effects through the PCFs. Further, hollow-core PCFs can be used to transmit high continuous-wave power with ultra-short pulses and very high peak powers [42]. The PCF has low-loss splices with effective cleaves; therefore, it is becoming more widely used in different applications, such as multiport couplers, and mode-area transformers. The PCF coupler can be simply realized using two defects through which the light propagates [43, 44]. The PCF coupler has better crosstalk, shorter coupling length, more design flexibility, and higher bandwidth values than a traditional coupler. The coupling characteristics of PCF couplers [43] are most important in optical fibers through which a division in wavelength or polarization can take place. The division process is also known as multiplexing–demultiplexing [44]. The PCF can also be used for polarization handling devices, such as polarization rotators [45–48], polarization splitters [49], and polarization filters [50].

2.8 Summary

This chapter discusses the fundamentals and types of PhCs and how the light is propagated and manipulated through the structure. Different configurations for PhC structures, PBG calculations, types of defects, and fabrication techniques are also described in detail. Additionally, the applications of the PhCs and PCFs are discussed.

References

[1] Joannopoulos, J.D., Meade, R.D. and Winn, J.N. (2001) *Photonic Crystals: Molding the Flow of Light*, Princeton University Press, Princeton.
[2] Morton, A. V. (2002) Photonic band structure calculations for 2D and 3D photonic crystals. Master thesis. San Jose State University.

[3] Yablonovitch, E. (May 1987) Inhibited spontaneous emission in solid state physics and electronics. *Phys. Rev. Lett.*, **58**, 2059.

[4] John, S. (1987) strong localization of photons in certain disordered dielectric superlattices. *Phys. Rev. Lett.*, **58**, 2486.

[5] Chen, Y.-C. (2009) Fabrication and characterization of three dimensional photonic crystals generated By multibeam interference lithography. PhD thesis. University of Illinois at Urbana-Champaign.

[6] Zhao, J., Li, X., Zhong, L. and Chen, G. (2009) Calculation of photonic band-gap of one dimensional photonic crystal. *J. Phys. Conf. Ser.*, **183**, 012018.

[7] Brzozowski, L. (2003) Optical signal processing using nonlinear periodic structures. PhD thesis. Toronto University.

[8] White, D. (June 2004) Photonic band-gap waveguides. Article, Lawrence Livermore National Laboratory, California.

[9] Qiu, M. (2000) Computational methods for the analysis and design of photonic band gap structures. PhD thesis. Division of Electromagnetic Theory, Department of Signals, Sensors, and Systems, Royal Institute of Technology, Stockholm, September.

[10] Lourtioz, J.M., Benisty, H., Berger, V., Gerard, J.M., Maystre, D. and Techelonkov, A. (2005) *Photonic Crystals Towards Nanoscale Photonic Device*, Springer Verlag, Berlin/Heidelberg.

[11] Kittel, C. (1971) *Introduction to Solid State Physics*, John Wiley & Sons, Inc., New York.

[12] Yablonovitch, E. (December 2001) Photonic crystals: semiconductors of light. *Sci. Am.*, **285**, 48–58.

[13] Johnson, S.G. and Joannopoulos, J.D. (2003) Introduction to Photonic Crystals: Bloch's Theorem, Band Diagrams, and Gaps (But No Defects), MIT Tutorial.

[14] Hermann, D., Frank, M., Busch, K. and Wolfle, P. (January 2001) Photonic band structure computations. *Opt. Express*, **8** (3), 167–172.

[15] Guo, Sh. (2001) *Plane Wave Expansion Method for Photonic Band Gap Calculation Using MATLAB, Manual Version 1.00*, Norfolk, Department of Electrical & Computer Engineering, Old Dominion University.

[16] Shumpert, J.D. (2001) Modeling of periodic dielectric structures (electromagnetic crystals). PhD thesis. Michigan University.

[17] Zheng, L.G. and Zhang, W.X. (2003) Study on bandwidth of 2-D dielectric PBG material. *Prog. Electromagn. Res.*, **41**, 83–106.

[18] Sibilia, C., Benson, T.M., Marciniak, M. and Szoplik, T. (eds) (2008) *Photonic Crystals: Physics and Technology*, Springer, Berlin.

[19] Lai, N.D., Lin, J.H., Do, D.B. *et al.* (2011) *Fabrication of Two- and Three-Dimensional Photonic Crystals and Photonic Quasi-Crystals by Interference Technique*, InTech, Croatia.

[20] Vogelaar, L. (2001) Large area photonic crystal slabs for visible light with waveguiding defect structures: fabrication with focused ion beam assisted laser interference lithography. *Adv. Mater.*, **13**, 1551.

[21] Vlasov, Y.A., Bo, X.-Z., Sturm, J.C. and Norris, D.J. (2001) On-chip natural assembly of silicon photonic bandgap crystals. *Nature*, **414**, 289.

[22] Wong, S., Kitaev, V. and Ozin, G.A. (November 2003) Colloidal crystal films: advances in universality and perfection. *J. Am. Chem. Soc.*, **125**, 15589–15598.

[23] Berger, V., Gauthier-Lafaye, O. and Costard, E. (1997) Photonic band gaps and holography. *J. Appl. Phys.*, **82**, 60–64.

[24] Campbell, M., Sharp, D.N., Harrison, M.T., Denning, R.G. and Turberfield, A.J. (March 2000) Fabrication of photonic crystals for the visible spectrum by holographic lithography. *Nature*, **404**, 53–56.

[25] Deubel, M., Freymann, G.V., Wegener, M., Pereira, S., Busch, K. and Soukoulis, C.M. (July 2004) Direct laser writing of three-dimensional photonic-crystal templates for telecommunications. *Nat. Mater.*, **3**, 444–447.

[26] HongboLan (2013) *Soft UV Nanoimprint Lithography and Its Applications*, InTech, Croatia.

[27] Mekis, A., Chen, J.C., Kurland, I., Fan, S.H., Villeneuve, P.R. and Joannopoulos, J.D. (1996) High transmission through sharp bends in photonic crystal waveguides. *Phys. Rev. Lett.*, **77**, 3787–3790.

[28] Takano, H., Akahane, Y., Asano, T. and Noda, S. (2004) In-plane-type channel drop filter in a two-dimensional photonic crystal slab. *Appl. Phys. Lett.*, **84**, 2226–2228.

[29] Chutinan, A., Mochizuki, M., Imada, M. and Noda, S. (2001) Surface-emitting channel drop filters using single defects in two-dimensional photonic crystal slabs. *Appl. Phys. Lett.*, **79**, 2690–2692.

[30] Akahane, Y., Asano, T., Song, B.S. and Noda, S. (2003) Investigation of high-Q channel drop filters using donor-type defects in two-dimensional photonic crystal slabs. *Appl. Phys. Lett.*, **83**, 1512–1514.

[31] Takano, H., Song, B.S., Asano, T. and Noda, S. (2006) Highly efficient multi-channel drop filter in a two-dimensional hetero photonic crystal. *Opt. Express*, **14**, 3491–3496.

[32] Okano, M., Kako, S. and Noda, S. (2003) Coupling between a point-defect cavity and a line-defect waveguide in three-dimensional photonic crystal. *Phys. Rev. B*, **68**, 235110.

[33] Campenhout, J.V., Bienstman, P. and Baet, R. (2005) Band-edge lasing in gold-clad photonic-crystal membranes. *IEEE J. Sel. Areas Commun.*, **23**, 1418–1423.

[34] Sakai, K., Miyai, E., Sakaguchi, T., Ohnishi, D., Okano, T. and Noda, S. (2005) Lasing band-edge identification for a surface emitting photonic crystal laser. *IEEE J. Sel. Areas Commun.*, **23**, 1335–1340.

[35] Knight, J.C., Birks, T.A. and Russell, P.S.J. (2001) Holey silica fibres, in *Optics of Nanostructured Materials*, vol. 39 (eds V.A. Markel and T.F. George), chapter 2, pp. 39–71, John Wiley & Sons, Inc., New York.

[36] Knight, J.C., Birks, T.A., Russell, P.S.J. and Atkin, D.M. (October 1996) All-silica single-mode fiber with photonic crystal cladding. *Opt. Lett.*, **21**, 1547–1549.

[37] Birks, T.A., Knight, J.C. and Russell, P.S.J. (1997) Endlessly single-mode photonic crystal fibre. *Opt. Lett.*, **22**, 961–963.

[38] Knight, J.C., Birks, T.A., Cregan, R.F., Russell, P.S.J. and de Sandro, J.P. (1998) Large mode area photonic crystal fiber. *Electron. Lett.*, **34**, 1347–1348.

[39] Gander, M.J., McBride, R., Jones, J.D.C. *et al.* (1998) Experimental measurement of group velocity in photonic crystal fiber. *Electron. Lett.*, **35**, 63–64.

[40] Cregan, R.F., Mangan, B.J., Knight, J.C. *et al.* (September 1999) Single-mode photonic band gap guidance of light in air. *Science*, **285** (5433), 1537–1539.

[41] Joshi, P., Sharma, R.K., Kishore, J. and Kher, S. (2007) Fabrication of photonic crystal fibre. *Curr. Sci.*, **93** (9), 1214–1215.

[42] Russell, P.S.J. (2006) Photonic-crystal fibers. *J. Lightwave Technol.*, **24** (12), 4729–4749.

[43] Mangan, B.J., Knight, J.C., Birks, T.A., Russell, P.S.J. and Greenaway, A.H. (2000) Experimental study of dual corephotonic crystal fiber. *Electron. Lett.*, **36**, 1358–1359.

[44] Saitoh, K., Sato, Y. and Koshiba, M. (2003) Coupling characteristics of dual-core photonic crystal fiber couplers. *Opt. Express*, **11**, 3188–3195.

[45] Hameed, M.F.O., Obayya, S.S.A. and El-Mikati, H.A. (February 2012) Passive polarization converters based on photonic crystal fiber with L-shaped core region. *J. Lightwave Technol.*, **30** (3), 283–289.

[46] Hameed, M.F.O. and Obayya, S.S.A. (August 2012) Design consideration of polarization converter based on silica photonic crystal fiber. *IEEE J. Quantum Electron.*, **48** (8), 1077–1084.

[47] Hameed, M.F.O., Abdelrazzak, M. and Obayya, S.S.A. (January 1, 2013) Novel design of ultra-compact triangular lattice silica photonic crystal polarization converter. *J. Lightwave Technol.*, **31** (1), 81–86.

[48] Hameed, M.F.O., Heikal, A.M. and Obayya, S.S.A. (August 15, 2013) Novel passive polarization rotator based on spiral photonic crystal fiber. *IEEE Photon. Technol. Lett.*, **25** (16), 1578–1581.

[49] Hameed, M.F.O. and Obayya, S.S.A. (December 2009) Polarization splitter based on soft glass nematic liquid crystal photonic crystal fiber. *IEEE Photon. J.*, **1** (6), 265–276.

[50] Mohamed, F.O., Hameed, A.M., Heikal, B.M., Younis, M.A. and Obayya, S.S.A. (2015) Ultra-high tunable liquid crystal-plasmonic photonic crystal fiber polarization filter. *Opt. Express*, **23**, 7007–7020.

3

Fundamentals of Liquid Crystals

3.1 Introduction

Liquid crystals (LCs) [1] are materials with an intermediate phase between the solid and fluid state. Liquids molecules, as shown in Figure 3.1, are mobile with no orientational or positional order, whereas the solid-state molecules have no mobility with long-range orientational and positional order. On the other hand, the LC molecules can be oriented as crystals with the ability to flow in a liquid-like way with a combination of both order and mobility. Therefore, LC phases are called intermediate phases or mesophases. There are different types of LC phases with different optical properties such as thermotropic, lyotropic, and metallotropic phases. The phase transition of the thermotropic LCs into the LC phase is temperature dependent. In addition, the thermotropic LCs will be converted into the conventional liquid phase at high temperature with random and isotropic molecular order. However, the phase transition of thelyotropic LCs depends on the temperature and concentration of the LC molecules in a solvent. Both thermotropic and lyotropic LCs are composed of organic molecules, while the metallotropic LCs have organic and inorganic molecules. Therefore, the LC phase transition of the metallotropic LCs is a function of temperature, concentration, and the inorganic–organic composition ratio.

In this chapter, the molecular structure and chemical composition of LCs, their classification, and their basic physical and optical properties are first introduced. Next, the theoretical treatments and sample preparation of LC cells are described. Applications of LCs for electro-optics and display-related applications are presented and new LC developments in photonics are drawn.

Computational Liquid Crystal Photonics: Fundamentals, Modelling and Applications, First Edition.
Salah Obayya, Mohamed Farhat O. Hameed and Nihal F.F. Areed.
© 2016 John Wiley & Sons, Ltd. Published 2016 by John Wiley & Sons, Ltd.

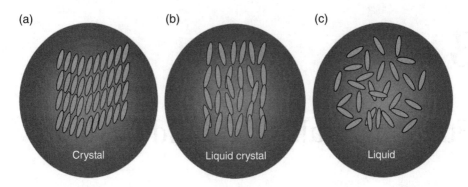

Figure 3.1 A schematic diagram of the structure of the (a) crystal, (b) liquid crystal, and (c) liquid states

Figure 3.2 Typical shape of an LC molecule

3.2 Molecular Structure and Chemical Composition of an LC Cell

The thermotropic LCs are usually formed by anisotropic-shaped molecules such as elongated rod- or disk-like. Figure 3.2 shows the molecular structure of a typical rod-like LC molecule. It consists of two or more ring systems connected by a central linkage group [2–4].

The rings can provide the short-range molecular forces needed to form the nematic phase. In addition, they can affect the electrical and elastic properties of the LC material. Moreover, the central linkage group has a great impact on the chemical stability of LCs, including their resistance to moisture or ultraviolet radiation. In this regard, the most stable LC compound has a single bond in the center. The elastic constants and the transition temperature of the LC phases are mainly controlled by the long side-chain at one side of the rings. However, the dielectric constant of the LC and its anisotropy are determined by the terminal group at the other end of the rings. Figure 3.3 shows some simple examples of molecules with an LC phase [3–6]. These examples include MBBA (4′-methoxybenzylidiene-4-butylaniline), 6CHBT (4-*trans-n*-hexyl-cyclohexyl-isothiocyanatobenzene), 5PCH (4-(*trans*-4′-pentyl-cyclohexyl)-benzonitril), and 5CB (4-pentyl-4′-cyanobiphenyl). In addition, there are more complicated types of LCs such as disk-like or banana-shaped [7] as shown in Figure 3.4. Moreover, chiral molecules without mirror symmetry can be arranged to obtain cholesteric LC phases.

3.3 LC Phases

The LC phases [1, 8] are mainly dependent on the type of ordering of the molecules. The molecules may be arranged in an ordered lattice, called positional order, and can have orientational order with specific direction. The positional order of the LC molecules may occur in a short range between molecules close to each other or in a long range. The LC phases can be classified into two main groups: thermotropic and lyotropic. Thermotropic LC phases are

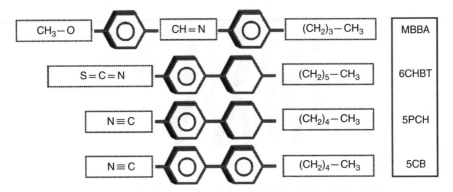

Figure 3.3   Examples of molecular structure of LC materials

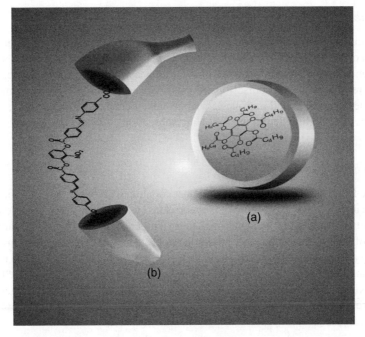

Figure 3.4   Molecular structure of (a) benzene-hexa-*n*-alkanoate derivatives and (b) banana-shaped LCs

obtained using organic molecules in a certain temperature range. Thus, lyotropic LC phases form a solution, and hence the liquid crystallinity is dependent on the concentration as well as the temperature. Therefore, lyotropic LCs are formed by amphiphilic molecules in solution, while thermotropic organic molecules do not need a solvent to form an LC phase. Thermotropic LCs attract the interest of many researchers all over the world due to their applications in electro-optic displays and sensing devices, while lyotropic LCs are important in biological applications. In the following subsections, the difference between the LC mesophases will be explained briefly.

(Liquid crystal)

Figure 3.5 Schematic diagram of nematic LC phase alignment

3.3.1 Thermotropic LCs

The formation of the thermotropic mesophase [1] is a function of the temperature and can occur in a certain temperature range. At high temperature, the LC phase ordering will be destroyed, and hence a conventional isotropic liquid phase will be formed, whereas at too low a temperature, a conventional anisotropic crystal will be obtained. There are many thermotropic LC mesophases, called mesogens, such as nematic, smectic, chiral, blue, and discotic phases that will be highlighted in the following subsections.

3.3.1.1 Nematic Phase

The nematic LC (NLC) phase is one of the most common LC phases at which the orientation of the molecules, as shown in Figure 3.5, is correlated with a preferred local orientation. The local orientation of the molecules is described by a unit vector pointing in the same direction as the axis of the LC molecules and called director n. The director's orientation can be controlled using an external static electric field. In this regard, the LC molecules tend to align their axis according to the applied field, and so the refractive indices of the NLCs are temperature dependent NLCs can also be tuned thermally. Because of these tunable properties, the NLCs have many related applications, especially in LC displays (LCDs).

3.3.1.2 Smectic Phase

The molecules of the smectic phase have positional order together with the orientational order. In addition, the molecules are arranged in layers with a well-defined layer periodicity. Smectic phases are obtained at lower temperature than nematic phases and are more ordered than nematics since at low temperatures, thermal vibrations of the molecules will be reduced and hence better ordering can be obtained. There are many different types of smectic mesophases with a variety of molecular arrangements within the layers. Figure 3.6 shows the most common smectic phases: A and C. Both smectic phases have no positional order between molecules within the layers. In addition, the molecules of smectic A phase, as shown in Figure 3.6a, are oriented along the layer normal, whereas the smectic C phase molecules

(a)   (b)

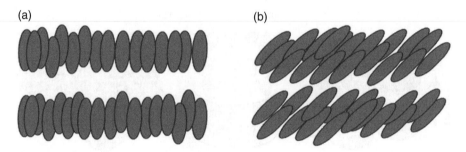

Figure 3.6   Schematic diagram of smectic LC phases alignment in (a) the smectic-A phase and (b) the smectic-C phase

Figure 3.7   Chiral nematic LC phase

are tilted away from the layer normal, as shown in Figure 3.6b. The layers of the smectic phases can slide over one another like soap. The smectics are thus positionally ordered along one direction.

3.3.1.3   Chiral Phases

The chiral nematic or cholesteric phase is the counterpart of the nematic phase. The chiral phase can be obtained either by chiral molecules in nature or by doping with small quantities of chiral molecules. Figure 3.7 shows the structure of the cholesteric LC phase. It is evident from the figure that the cholesteric LC phase consists of quasi-nematic layers. Each layer has an individual director that is turned by a fixed angle on moving from one layer to the next. The distance between the layers required for the director of the molecule to rotate a full 360° is called the hole pitch, p as shown in Figure 3.7. Therefore, the director describes a helix with a specific temperature-dependent pitch. It should be noted that the structure of the chiral nematic phase repeats itself every $p/2$, because the directors at 0° and ±180° are equivalent. In addition, the pitch, p, can be tuned thermally or by doping the LC host with chiral material. Moreover, the pitch p may be of the same order as the wavelength of visible light. This offers uncommon optical properties, such as Bragg reflection, which can be used for optical applications [9].

(a)   (b)   (c)

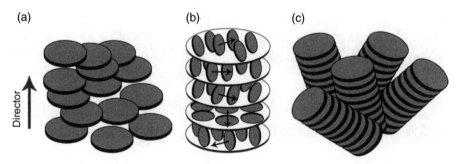

Figure 3.8 Schematic diagram of different types of discotic liquid crystalline phases, (a) nematic, (b) chiral nematic, and (c) nematic columnar

3.3.1.4   Blue Phases

Blue phases are LC phases that exist within the temperature range between a chiral nematic phase and a conventional isotropic liquid phase. The LC blue phases have regular 3D cubic defect structures with lattice periods of the same order as the wavelength of visible light . Therefore, they exhibit selective Bragg reflections in the visible wavelength range corresponding to the cubic lattice. Consequently, the LC blue phases can be used for useful applications such as fast light modulation and tunable photonic bandgap structures. However, blue phases have been found over a very small range in temperature, usually less than a few Kelvins. As a result, the stability of the LC blue phases is under intense research. In this regard, a new class of blue-phase LCs that remain stable over a temperature range from 260 to 326 K including room temperature has been demonstrated [10, 11].

3.3.1.5   Discotic Phases

This type of phase depends on the molecular shape which is a disk-shaped LC molecule. These molecules can be organized in columns to form columnar LC phases, which may be organized into rectangular, square, or hexagonal arrays. The organization of the disk-shaped molecules results in discotic mesophases [12]. In addition, the molecules of the discotic meso-phase are called discotic mesogens. Figure 3.8 shows schematic diagrams of different types of discotic NLC phases such as nematic, chiral nematic, and nematic columnar.

It can been seen from Figure 3.8a that the molecules of the discotic nematic phase are approximately parallel with only orientational order and short-range positional order. In addition, the disk-shaped molecules exhibit full translational and rotational freedom around their short axes. However, their long axes are oriented parallel to a general plane. On the other hand, the chiral discotic nematic mesophases, as shown in Figure 3.8b, result from mixtures of discotic nematic, non-mesogenic chiral dopants, and pure chiral discotic molecules. Moreover, a nematic columnar phase, as shown in Figure 3.8c, is characterized by a columnar stacking of the discotic molecules. In this case, the columns do not constitute two-dimensional (2D) lattice structures with a positional short-range order and an orientational long-range order.

Figure 3.9 shows schematic diagrams of columnar phases formed by discotic mesogens such as hexagonal, rectangular, oblique, and lamellar. It is evident that the molecules of the discotic columnar phase are formed in columns arranged in a 2D lattice. In addition, the columnar

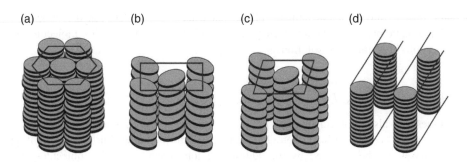

Figure 3.9 Schematic diagram of different columnar phases formed by discotic mesogens: (a) hexagonal, (b) rectangular, (c) oblique, and (d) lamellar

mesophases can be classified into different types such as hexagonal, rectangular, oblique, and lamellar, as shown in Figure 3.9. This depends on the 2D lattice arrangement and the molecular stacking order through the columns. For example, the molecular columns are arranged in hexagonal and rectangular lattices in the columnar hexagonal and rectangular phases, as shown in Figure 3.8a and b, respectively. However, in the columnar oblique phase, the columns are organized in an oblique unit cell, as can be seen in Figure 3.9c. Additionally, the molecules of the columnar lamellar mesophase are stacked into columns arranged in a layered structure.

3.3.2   Lyotropic LCs

A lyotropic LC can offer ordered mesophases in solutions rather than in their pure form. Therefore, lyotropic LCs consist of two or more components in certain concentration ranges. In this regard, the formation of the mesophase depends on the concentration of the mesomorphic material in the solution as well as its temperature. Therefore, various phases for the lyotropics can be obtained at different concentrations and temperatures. The molecules with two immiscible hydrophilic and hydrophobic parts that form the lyotropic LC system are called amphiphilic molecules. There are many amphiphilic molecules that can form lyotropic LC phases based on the volume balances between the hydrophilic and hydrophobic parts.

The amphiphilic molecules can form micelles with circular or rod-shaped micelles in aqueous solutions [13]. The rod-shaped micelles can be arranged in a hexagonal lattice through the solution, which results in a hexagonal lyotropic LC phase. This type of phase is highly viscous due to the high concentration of LC molecules. However, the circular micelles form cubic lyotropic LCs that are more viscous than the hexagonal phases due to the absence of shear planes. A lamellar lyotropic LC phase is formed at very high concentrations with water sandwiched between sets of two parallel sheets of molecules. The hydrophobic tails of the molecules in adjacent sheets point toward each other. In addition, the polar head groups point toward the water sandwiched between the two sheets.

The lyotropic LCs can also form nematic and cholesteric phases [13, 14]. Solutions containing disc-like micelles form lamellar nematics, while solutions with rod-like micelles form cylinder nematics. The change in concentration and/or temperature can convert these nematic phases into a biaxial nematic phase that usually exists between the lamellar and cylinder nematic phases. The cholesteric phases are obtained by using chiral molecules or by adding a chiral dopant to the solution.

3.3.3   Metallotropic LCs

LC phases can also be formed using low-melting inorganic phases such as $ZnCl_2$. These inorganic phases consist of linked tetra hedra and easily form glasses. Further, different types of LC phases can be obtained by the addition of long chain soap-like molecules. These new materials are called metallotropic [15] and are dependent on the inorganic–organic composition ratio and on the temperature.

3.4   LC Physical Properties in External Fields

LCs can be used in many applications due to the tunable control of the macroscopic properties of the LC system. In this regard, electric and magnetic fields can be used to control these properties. In addition, the magnitudes of the external fields as well as the corresponding arrangement of the LC director are the main factors for the choice of suitable applications. Additionally, special surface treatments and alignment layers can be used to enforce specific orientations of the director of the LC. In the following subsections, the effects of external fields on the LC orientation will be described briefly.

3.4.1   Electric Field Effect

The response of LC molecules to the application of an electric field depends on the dielectric properties of the LC. Assuming uniaxial LC phases with a macroscopic coordinate system x, y, z. If the z-axis is parallel to the director n of the LC, two principal permittivities can be obtained: $\varepsilon_\parallel$ and $\varepsilon_\perp$. $\varepsilon_\parallel$ is parallel to the director $\varepsilon_\parallel = \varepsilon_{zz}$, while $\varepsilon_\perp$ is perpendicular to the director where $\varepsilon_\perp = 1/2(\varepsilon_{xx} + \varepsilon_{yy})$. Therefore, the dielectric anisotropy of the LC material can be expressed as $\Delta\varepsilon = (\varepsilon_\parallel - \varepsilon_\perp)$, which may have positive or negative values. It should be noted that the permittivity of the material is a measure of how an electric field affects, and is affected by, a dielectric medium. Additionally, the permittivity is a measure of how easily a dielectric polarizes in response to an electric field.

The LC materials may have polar or nonpolar molecules. The polar molecules have permanent dipole moments due to the slight separation between positive and negative charges in the molecule. However, the induced electric dipoles through nonpolar molecules can be created by an applied electric field that can slightly separate the positive and negative charges in the molecules. In this case, the induced electric dipoles in the nonpolar molecules are much weaker than the permanent electric dipoles in the polar molecules; however, they experience the same forces in an electric field.

The permanent or induced dipoles occur through the LC molecules either along or across the long molecular axis. If the dipole moment and the long molecular axis are parallel, then $\Delta\varepsilon > 0$. Hence, the molecules will be oriented along the electric field direction, as shown in Figure 3.10a. However, the molecules will be oriented perpendicular to the electric field

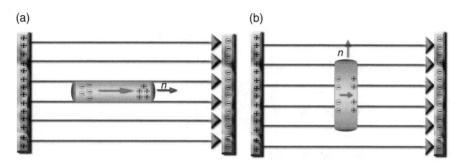

Figure 3.10 The effect of an electric field on LC molecules with (a) positive $\Delta\varepsilon$ and (b) negative $\Delta\varepsilon$

direction if the molecules have dipole moments that are normal to the long molecular axis with $\Delta\varepsilon < 0$, as shown in Figure 3.10b.

3.4.2 Magnetic Field Effect

The effect of magnetic fields on LC molecules is similar to that of the electric fields. Most LC materials consist of organic molecules that are diamagnetic. Therefore, permanent magnetic dipoles are produced by electrons moving about atoms. When a magnetic field is applied, the molecules will tend to align either with or against the field.

3.4.2.1 Frederiks Transition

The transition of a uniform director pattern in an external field to a deformed director configuration is called Frederiks transition. If an LC with uniform director is exposed to an external electric or magnetic field, a gradual change occurs through the director pattern when the field strength exceeds a threshold value. This means that the director pattern remains undistorted below the threshold field value. If the field value is gradually increased to greater than the threshold value, the director begins to twist until it is aligned with the field. In this regard, the Frederiks transition can occur in three different configurations: the twist, bend, and splay geometries. The threshold of the electric field value is given by

$$E_{th} = \frac{\pi}{d}\sqrt{\frac{K_i}{\varepsilon_0 \Delta\varepsilon}}, \tag{3.1}$$

where d is the separation distance between the two electrodes, K_i is an elastic constant, and $i = 1$, 2, and 3 correspond to splay, twist, and bend deformations, respectively. However, the threshold of the magnetic field value B_{th} is expressed as follows:

$$B_{th} = \frac{\pi}{d}\sqrt{\frac{K_i}{\mu_0^{-1}\Delta\chi}}. \tag{3.2}$$

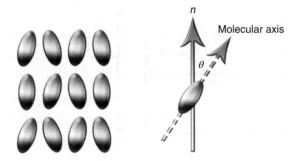

Figure 3.11 Molecular axis and director of a nematic LC

Here, μ_o is the magnetic permeability of vacuum, $\Delta\chi = \chi_\parallel - \chi_\perp$ is the diamagnetic anisotropy, $\chi_\parallel$ and $\chi_\perp$ are diamagnetic susceptibilities measured parallel and perpendicular to the director, respectively.

3.5   Theortitcal Tratment of LC

Theoretical descriptions of different LC phases are very important for investigating the connections between their molecular and macroscopic properties. The mathematical treatments of LCs are quite complicated owing to their properties of high material density and anisotropy. There are few simple theories that can be applied for predicting the general behavior of the phase transitions in different liquid crystalline states. In the following subsections, the molecular statistical treatments will be described for the simplest nematic crystalline state.

3.5.1   LC Parameters

The description of the LCs involves the analysis of two main parameters: director and order [16, 17].

3.5.1.1   Director

Figure 3.11 shows the NLCs that are composed of rod-like molecules. It can be seen from the figure that long molecules align such that they are on average parallel to a particular direction specified by a dimensionless unit vector n called the director or optical axis.

3.5.1.2   Order Parameter

This is the second parameter that can be used to characterize the LC structures. There are typically three types of orders that can be used to mathematically describe the LCs: orientational, positional, and bond orientational. Positional order describes the degree to which the positions of molecules exhibit translational symmetry. However, bond orientational order represents the line joining the centers of closest neighbor molecules without requiring a regular spacing along that line.

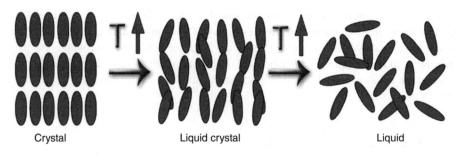

Figure 3.12   Effect of temperature on the order of LCs

The material may be more liquid-like or more crystal-like based on the degree of orientation and positional order in the LC phases. The orientation order of various LCs can be changed by increasing or decreasing the energy of the LC molecules. The orientational order parameter is a function of the average angle of displacement of the moment of inertia from the director and is usually defined based on the average of the second Legendre polynomial. The order is given by [16–18]

$$S = \left\langle \frac{3\cos^2 \theta - 1}{2} \right\rangle,\tag{3.3}$$

where θ is the angle between any molecule axis and the average director. The equation shows that a high-ordered LC is obtained by decreasing the average angle θ of displacement of the moment of inertia from the director. Thus, when θ is equal to zero for all molecules, the orientation parameter will be equal to 1 with a very high degree of organization order. Figure 3.12 shows that the orientational order decreases with increasing the temperature and then goes to zero at the clearing temperature at which the transition to an isotropic phase occurs. The orientational order parameter of most LCs varies between 0.3 and 0.9. However, the orientational order parameter of most of NLCs varies between 0 and 1.0.

3.5.2   LC Models

The scope of this section is to discuss a number of fairly simple models that can be applied to predict the general behavior of the phase transitions in LC structures.

3.5.2.1   Onsager Hard-Rod Model

This was proposed by Lars Onsager [17] to investigate approximately the lyotropic phase transitions based on several assumptions that limit its applicability to real systems. In this formalism, the volume excluded from the center of mass of one idealized cylinder as it approaches another is considered. More specifically, if the cylinders are aligned parallel to one another, there is too little volume that is excluded from the center of mass of the close cylinder. However, if the cylinders are at some angle to one another, then there is a large volume surrounding the cylinder which the approaching cylinder's center of mass cannot enter because of the hard-rod repulsion between the two idealized cylinders. Therefore, this angular arrangement shows a decrease in the net positional entropy of the approaching cylinder; and thereby, there are fewer states available to it. This model is conceptually helpful where the parallel arrangements of anisotropic objects minimize the orientational entropy. Therefore, this model shows that the solution of rod-shaped objects with suitable nematic phase concentration will undergo a phase transition.

3.5.2.2 Maier–Saupe Mean Field Theory

Wilhelm Maier and Alfred Saupe have proposed a microscopic mean field theory [6] for predicting the nematic–isotropic phase transitions. In the formalism of the proposed model Saupe *et al.* consider rod-like nonpolar molecules interacting via van der Waals interaction. Further, they consider the mean field approximation and assume that the orientation-dependent interaction has no impact on the configuration of the centers of the mass. The Maier–Saupe theory predicts the stable mesophase and the first-order phase transition as a function of temperature.

3.5.2.3 McMillan's Model

McMillan [19] has developed the Maier–Saupe mean field model to describe the phase transition of an LC from a nematic to a smectic-A phase. This developed model predicts that the phase transition can be either continuous or discontinuous depending on the strength of the short-range interaction between the molecules.

3.6 LC Sample Preparation

The crystalline properties of LCs are very similar to that of liquids. These properties in bulk, "unaligned" states become apparent by placing LC materials in flat thin cells. The alignment of the LC director in such cells is mainly dependent on the cell walls, where cell's surfaces are treated in various ways to obtain different director orientations [20].

There are many types of director axis alignments that are used for NLCs, such as homogeneous and homeotropic hybrid, twisted, supertwisted, fingerprint, and multidomain vertically aligned. However, the most common standard cell alignments are the homogeneous (planar) and homeotropic alignment techniques. In the homogeneous cell alignment, the long axis of the LC molecules is aligned along the plane of the cell wall. Alternatively, in the homeotropic alignment, the long axis of the LC molecules is aligned perpendicular to the cell wall plane. There is a common way of creating planar alignment by first coating the cell wall with some polymer-like polyvinyl alcohol (PVA), and then rubbing it in only one direction with a lens tissue. This operation introduces elongated pressure on the polymer and simplifies the planar alignment [20]. The following seven steps are used in the laboratory to prepare an NLC sample with a quite reliable homogeneous alignment:

1. Dissolve chemically pure PVA in distilled deionized water at an elevated temperature (close to the boiling point) at a concentration of about 0.2%.
2. At 300 K, dip two refined glass slides into the PVA solution and withdraw them bit by bit. This step results in growing film of the solution on each of the glass slides.

3. Dry the glass slides in an oven.
4. Rub the surface of the glass slides in one direction with a lens tissue.
5. Set up a nonreactive plastic spacer of appropriate dimension and place it on one of the glass slides.
6. Insert the NLC material into the inner spacer.
7. Put the second glass slide over the NLC and clamp the two slides together. Once gathered, the sample should be left alone for few minutes, and hence NLC sample with clear homogenous aligned state is obtained.

Alternatively, to obtain a quick and an effective homeotropic alignment, the cell walls should be dealt with soap surfactants as hexadecyl-trimethylammoniumbromide (HTAB). The molecules of these surfactants align themselves perpendicularly to the cell wall. The following seven steps are used in the laboratory to prepare a homeotropic alignment for the NLC sample:

1. Dissolve one fraction of HTAB in 50 fractions of distilled deionized water
2. Purify two glass slides.
3. Dip the two cleaned glass slides in the HTAB solution and withdraw them gradually. This results in leaving a film of HTAB molecules effectively on each of the glass slides.
4. Dry the glass slides in an oven.
5. Set up a nonreactive plastic spacer of suitable dimensions and put the spacer on one of the slides.
6. Insert the NLC material into the inner spacer.
7. Place the second glass slide over the NLC and clamp the two glass slides together. Once gathered, leave the sample alone for few minutes, and hence a sample of clear homeotropically aligned state is obtained.

In order to extend the sample life time, the sample preparation should be done in a refined room free of humidity or other chemicals. Generally, preparing NLCs materials such as 5CB and E7 using the techniques introduced formerly are quite stable against temperature and have been shown to last several months, whereas the performance of MBAA decays in a few days.

The previously outlined steps for preparing planar cell alignment can be applied effectively for chiral nematic LCs. Further, the previously introduced steps for preparing homeotropic cell alignment can be applied for smectic-A preparation. As the sample slowly cools down from the nematic to the smectic phase, an external field should be applied to help preserve the homeotropic alignment. The cell preparation methods for LCs such as ferroelectric LC (FLC) and smectic-C* are more complex as they involve surface stabilization. Alternatively, the cell preparation methods for other LCs such as smectic-A* (Sm-A*) cells are easier than the methods discussed before.

3.7 LCs for Display Applications

LCD technology is considered as the most common application of LC [21]. The LCD industry gains increasing importance and several significant scientific and engineering discoveries ranging from simple calculators, wrist watches, advanced VGA computer screens to full color LCD televisions have been made. The basic idea to present information on all LCD displays is to use a sandwiched LC material between two glass plates with ambient light for display illumination. This results in reducing the size of display, consuming much less power than cathode ray tube devices and making LCDs practical for applications where size and weight are important. However, LCDs are adopted in many applications, but they have drawbacks with viewing angle, contrast ratio, and response time.

3.8 LC Thermometers

Chiral nematic or cholesteric LCs can be applied effectively for implementing thermometer devices [22], based on the property of chiral NLCs for reflecting specific colored light based upon the temperature. Therefore, an LC makes it possible to accurately gauge temperature just by looking at the color of the thermometer. Thermometers can be easily built for any temperature range by mixing different compounds. Further, LC temperature sensors can be employed to discover bad connections on a circuit board and can also be attached to the skin to show a "map" of temperatures.

3.9 Optical Imaging

Different types of LCs are now being investigated and researched for optical imaging and recording [23]. In this promising technology, an LC cell is sandwiched between two layers of photoconductor. By applying light to the photoconductor, the material's conductivity increases and an electric field is developed in the LC based on the light intensity. Therefore, an image can be recorded by transmitting an electric pattern by an electrode.

3.10 LC into Fiber Optics and LC Planar Photonic Crystal

In addition to LC bulk thin film applications discussed in the previous sections, LCs have a multitude of other applications into optical waveguides and nanostructured photonic crystals [20–24]. Recently, it has been shown that LCs can be used for filling hollow fibers (microcapillaries) made of a material with a lower refractive index, such as Pyrex or silica glass, with the purpose of making easily tuned attenuators, equalizers, polarization controllers, phase emulators, nonlinear, high-quality image transmitting fiber arrays, and so on. It was reported that such structures can be easily fabricated and also show small scattering losses of about 3 dB/cm compared to 20 dB/cm for a slab waveguide [24]. Moreover, LC materials integrated with fiber have been employed for various sensing applications, such as chemical sensing, gas sensing [25], and biological sensing [26]. Fiber optic sensors [27] have several advantages, such as immunity from electromagnetic interference, integrity in possibly dangerous environments, and possibility of large separation among the sensors and monitoring stations. Additionally, LCs have been employed with photonic crystals in 1D, 2D, and 3D forms to obtain tunable filters, switches, splitters, mirrors and lasing devices with lower power consumption and lower insertion losses. Whinnery et al. and Giallorenzi et al. (see Ref. [9]) have proposed electro-optical and integrated optical switching devices based on placing a thin film (~1 μm) of LC between two glass slides that have lower refractive index than the LC. One of the two glass slides has an organic film deposited on it in order to receive the input laser that excites the transverse electric (TE) and/or transverse magnetic (TM) modes in the organic film, and is then guided into the NLC region. Though these are drawbacks with nematics, such as the large losses of about 20 dB/cm and relatively slow responses, they may be useful in various nonlinear optical applications [26].

3.11   LC Solar Cell

Solar energy is the most promising source for green sustainable renewable electric energy. The conversion of light to electric or chemical energy using photovoltaic technology has been known since 1839 and is based on light-induced electron–hole generation. Inorganic semiconductors are used to implement high-efficiency solar cells. Recently, organic thin-film solar cells have become popular for flat-plate terrestrial applications since they can be fabricated effectively using a large area substrate in an economical way. The two types of organic semiconductors Schottky and p/n-type are employed for fabricating photovoltaic solar cells. In the Schottky solar cell, the organic p-conductor is usually placed between two electrodes with different work functions, while in an organic double-layer p/n-solar cell layers of n and p-conductors are placed between two electrodes. Though these organic photovoltaic solar cells have been adopted, they exhibit very low power conversion efficiencies (<1%). Recent investigations show that blending donor and acceptor molecules can be utilized to optimize photo-induced charge separation, and thereby improve the conversion efficiency of the photovoltaic device. Shaheen *et al.* [28] recently improved the conversion efficiency into 1.5% by blending a soluble methanol fullerene into the conjugated polymer film. In addition, the doping of bromine molecules to an organic–inorganic heterojunction device based on ZnO and single crystalline pentacene produces a solar cell with conversion efficiencies up to 4.5%. Though the crystalline molecular organic materials exhibit better conversion efficiency, the processing of single crystals is difficult. This obstacle can be overcome by using discotic LCs since their columnar structure resembles the aromatic stacking in single crystalline conductors. Recently, Schmidt-Mende *et al.* [29] reported an efficient organic photovoltaic solar cell designed using hexabenzocoronene-based discotic LCs. In addition, Gregg *et al.* [30] proposed and analyzed symmetrical photovoltic cells filled with discotic liquid crystalline porphyrin. As discovered by [30], there was a similarity between the photovoltaic effect of the proposed structure with that of the better organic solar cells. Further, Schmidt-Mende *et al.* [29] introduced a p/n-type photovoltaic solar cell utilizing discotic liquid crystalline hexa benzocoronene as the hole-transporting layer and a perylene dye as electron transporting layer. The device made by Schmidt-Mende *et al.* [29] exhibits power efficiencies of up to ~ 2%. The main insight of the proposed solar cell by Schmidt-Mende *et al.* [29] is due to simple solution processing and utilization of the self-organization of discotic liquid crystalline and crystalline organic materials for the charge transport. These promising results with the aid of several other recent studies on discotic LCs pave the way to achieving cheaper, clean, efficient green energy.

References

[1] de Gennes, P.G. and Prost, J. (1995) *The Physics of Liquid Crystals* (2nd ed.), Clarendon Press, Oxford Science Publications, Oxford.

[2] Khoo, I.-C. (ed) (1994) *Liquid Crystals: Physical Properties and Nonlinear Optical Phenomena*, Wiley-Interscience, New York.

[3]   Yeh, P. and GU, C. (eds) (1999) *Optics of Liquid Crystal Displays*, Wiley-Interscience, New York.

[4]   Desimpel, C. (2005–2006), Liquid crystal devices with in-plane director rotation. PhD thesis. Electronics and Information Systems, Gent.

[5]   Blinov, L.M. and Chigrinov, V.G. (eds) (1996) *Electrooptic Effects in Liquid Crystal Materials, Partially Ordered Systems*, Springer-Verlag, New York.

[6]   Sarkar, P., Mandal, P.K., Paul, S., and Czuprynski, K. (2003) X-ray diffraction, optical birefringence, dielectric and phase transition properties of the long homologous series of nematogens 4-(trans-4'-n-alkylcyclohexyl) isothiocyanatobenzenes. *Liq. Cryst.*, **30**, 507–527.

[7]   Senyuk, B. *Liquid Crystals: A Simple View on a Complex Matter*, http://www.personal.kent.edu/~bisenyuk/ liquidcrystals/index.html (accessed January 4, 2016).

[8]   Chandrasekhar, S. (1992) *Liquid Crystals*, 2nd edn, Cambridge University Press, Cambridge.

[9]   Sluckin, T.J., Dunmur, D.A., and Stegemeyer, H. (eds) (2004) *Crystals That Flow: Classic Papers from the History of Liquid Crystals*, Taylor & Francis, London.

[10] Coles, H.J. and Pivnenko, M.N. (2005) Liquid crystal/"blue phases/" with a wide temperature range. *Nature*, **436**, 997–1000.

[11] Yamamoto, J., Nishiyama, I., Inoue, M., and Yokoyama, H. (2005) Optical isotropy and iridescence in a smectic /'blue phase/'. *Nature*, **437**, 525–528.

[12] Kumar, S. (2006) Self-organization of disc-like molecules: chemical aspects. *Chem. Soc. Rev.*, **35**, 83–109.

[13] Tamhane, K. (2009) Formation of lyotropic liquid crystals through the self-assembly of bile acid building blocks. Master of Science. Mechanical, Materials and Aerospace Engineering University of Central Florida, Florida.

[14] Collings, P. and Patel, J.S. (eds) (1997) *Handbook of Liquid Crystal Research*, Oxford University Press, New York.

[15] Martin, J.D., Keary, C.L., Thornton, T.A., Novotnak, M.P., Knutson, J.W., and Folmer, J.C.W. (2006) Metallotropic liquid crystals formed by surfactant templating of molten metal halides. *Nat. Mater.*, **5**, 271–275.

[16] Dierking, I. (2003) *Textures of Liquid Crystals*, Wiley-VCH, Weinheim.

[17] Onsager, L. (1949) The effects of shape on the interaction of colloidal particles. *Ann. N. Y. Acad. Sci.*, **51**, 627.

[18] Collings, P.J. and Hird, M. (1997) *Introduction to Liquid Crystals*, Taylor & Francis, Bristol.

[19] McMillan, W. (1971) Simple molecular model for the smectic A phase of liquid crystals. *Phys. Rev. A*, **4** (3), 1238.

[20] Khoo, I.-C. (2007) *Liquid Crystal*, 2nd edn, Wiley-Interscience, Hoboken.

[21] Castellano, J.A. (2005) *Liquid Gold: The Story of Liquid Crystal Displays and the Creation of an Industry*, World Scientific Publishing, Hackensack.

[22] Harris, W.F. (1978) Dislocations, disclinations and dispirations: distractions in very naughty crystals. *S. Afr. J. Sci.*, **74**, 332.

[23] Andrienko, D. (September 14, 2006) *Introduction to Liquid Crystals, Lecture Notes*, http://www2.mpipmainz. mpg.de/~andrienk/lectures/IMPRS/liquid_crystals.pdf (accessed January 4, 2016).

[24] Dutton, H. J. R. (September 1998) Understanding optical communication, international technical organization support.

[25] Shah, R.R. and Abbott, N.L. (2001) Principles for measurement of chemical exposure based on recognition-driven anchoring transitions in liquid crystals. *Science*, **293**, 1296–1299.

[26] Zou, Y., Namkung, J., Lin, Y., and Lindquist, R. (2010) *Enhanced chemical and biological sensor based on liquid crystal using a bias electric field*. Proceedings of the Laser and Electro-optics, Quantum Electronics and Laser Science, May 2010, San Jose, CA.

[27] Mathews, S. (2011) Liquid crystal devices for optical communications and sensing applications. PhD thesis. Dublin Institute of Technology.

[28] Shaheen, S.E., Brabec, C.J., Sariciftci, N.S., Padinger, F., Fromherz, T. and Hummelen, J.C. (2001) 2.5% efficient organic plastic solar cells. *Appl. Phys. Lett.*, **78**, 841–843.

[29] Schmidt-Mende, L., Fechtenkotter, A., Mullen, K., Moons, E., Friend, R.H. and Mackenzie, J.D. (2001) Self-organized discotic liquid crystals for high-efficiency organic photovoltaic's. *Science*, **293**, 1119.

[30] Gregg, B.A., Fox, M.A. and Bard, A.J. (1990) Photovoltaic effect in symmetrical cells of a liquid crystal porphyrin. *J. Phys. Chem.*, **94**, 1586.

Part II

Numerical Techniques

4

Full-Vectorial Finite-Difference Method

4.1 Introduction

During the past decade, various accurate modeling methods have been developed for modal analysis of different photonic devices. These methods included the finite-difference method (FDM) [1], the finite element method (FEM) [2], and the multipole method [3]. However, the modal solution techniques based on the FDM are very popular due to their simple implementations. An overview of the modeling methods is presented in this chapter. In addition, the formulation of the FDM [1] with perfect boundary conditions capabilities is explained in detail.

4.2 Overview of Modeling Methods

Numerical simulations play a fundamental role in the design and modeling of photonic crystal devices, such as photonic crystal fiber (PCF). However, various modeling methods, either full-vector models or approximate scalar models, have been developed. An approximate scalar model [4] is a worthy tool for facilitating the fabrication because it is easy to use and provides good qualitative information. However, in order to model PCFs accurately and investigate sensitive quantities such as dispersion and birefringence, full-vector models should be applied. The FEM [3] can provide high accuracy by means of flexible triangular and curvilinear meshes to represent the waveguide cross section. However, this accuracy results in an algorithm that is complex to implement. Also, the finite element-based imaginary distance beam propagation method (IDBPM) has also been suggested as a "mode solver" for PCFs [5]. Although accurate and versatile, IDBPM may sometimes lack convergence if a suitable initial field distribution is not chosen. Moreover, if higher order modes are needed, successive running of the iterative

Computational Liquid Crystal Photonics: Fundamentals, Modelling and Applications, First Edition.
Salah Obayya, Mohamed Farhat O. Hameed and Nihal F.F. Areed.
© 2016 John Wiley & Sons, Ltd. Published 2016 by John Wiley & Sons, Ltd.

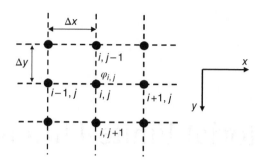

Figure 4.1   A part of a typical uniform finite-difference mesh

IDBPM code is required, which is not computationally efficient. On the contrary, the mode solvers based on the FDM [1] and multipole method [3] are very attractive because of their simple implementations. The FDM is probably the most widely used technique for the numerical modeling of PCF waveguides. The appropriate algorithms for a rough finite-difference solution to a structure with a very general geometry can be programmed very quickly. At the same time, FDMs offer a degree of flexibility with nonuniform meshing and are also applicable to all of the common modeling problems, including modal and propagation analyses.

The FDM is a procedure for transforming a partial differential equation into a finite set of linear equations. To do this, a rectangular mesh composed of lines parallel to the coordinate axis is superimposed on the problem space. The mesh points lie at the intersection of the mesh lines and form the discrete set of points at which the function values are stored. A portion of a two-dimensional finite-difference mesh is shown in Figure 4.1. In the figure, the mesh lengths, Δx and Δy, are the distances between any two adjacent mesh points, in the x and y directions, respectively. For each mesh point, i and j with corresponding coordinates x_i and y_j, an approximate expression (or difference equation) $\varphi_{i,j} = \varphi(x_i, y_j)$ for the partial differential equation is formed that involves the function value at i, j, and at some of the surrounding points. This is normally done by expanding the field at various neighboring points as a Taylor series to produce expressions for the derivatives occurring in the partial differential equation at i and j. Once this process has been performed for each point in the problem space, the complete set of linear equations relating the function values at all points is solved to give an approximate solution to the partial differential equation.

4.3   Formulation of the FVFDM

4.3.1   Maxwell's Equations

In the frequency domain, the propagation of electromagnetic waves through the waveguide is governed by a set of four partial differential Maxwell's equations that can be written as follows:

$$\nabla \times E = -j\omega\mu H \tag{4.1}$$

$$\nabla \times H = j\omega\bar{\bar{\varepsilon}} E \tag{4.2}$$

$$\nabla \cdot \left(\bar{\bar{\varepsilon}} E \right) = 0 \tag{4.3}$$

$$\nabla \cdot \mu H = 0. \tag{4.4}$$

Here, the vector quantities E and H are electric and magnetic field vectors, respectively, $\bar{\bar{\varepsilon}} = \varepsilon_0 \bar{\bar{\varepsilon}}_r$ and $\mu = \mu_0 \mu_r$. The quantities $\bar{\bar{\varepsilon}}$ and μ define the electromagnetic properties of the medium and are the permittivity tensor and the permeability of the waveguide material, respectively. $\varepsilon_0 = 8.854 \times 10^{-12}$ F/m is the permittivity of free space, and $\mu_0 = 4\pi \times 10^{-7}$ H/m is the permeability of free space. $\bar{\bar{\varepsilon}}_r$ and $d\mu_r$ are the relative permittivity tensor and the permeability of the waveguide material. In this study for nonmagnetic waveguides, μ_r is taken as unity.

4.3.2 Wave Equation

The wave equation can be expressed in terms of only the electric or the magnetic field components. In this section, the derivation that models the magnetic field is considered. In this case, the electric field is removed from the derivation by taking the curl of Eq. (4.2) and substituting using (4.1). The vector wave equation for the magnetic field vector, H, can be written as follows:

$$\nabla \times \left(\bar{\bar{\varepsilon}}_r^{-1} \nabla \times H \right) - k_o^2 H = 0. \tag{4.5}$$

Here, k_o is the free space wave number and $k_o^2 = \omega^2 \mu_0 \varepsilon_0$. The anisotropic material is assumed to have one of its principal axes pointing in the direction of the waveguide. In this way, the permittivity tensor takes the following form:

$$\bar{\bar{\varepsilon}} = \varepsilon_0 \bar{\bar{\varepsilon}}_r = \varepsilon_0 \begin{pmatrix} \varepsilon_{xx} & \varepsilon_{xy} & 0 \\ \varepsilon_{yx} & \varepsilon_{yy} & 0 \\ 0 & 0 & \varepsilon_{zz} \end{pmatrix} \tag{4.6}$$

For isotropic waveguides, $\varepsilon_{xx} = \varepsilon_{yy} = \varepsilon_{zz}$ and $\varepsilon_{xy} = \varepsilon_{yx} = 0$. Using the vector notation $(\nabla \times (\varphi A) = \nabla \varphi \times A + \varphi \nabla \times A)$, the vector wave equation (4.5) can be rewritten as follows:

$$\nabla^2 H + k_o^2 \bar{\bar{\varepsilon}}_r H = -\bar{\bar{\varepsilon}}_r^{-1} \nabla \bar{\bar{\varepsilon}}_r \times (\nabla \times H). \tag{4.7}$$

The transverse component of the vector wave equation (4.7) can be expressed as follows:

$$\nabla^2 H_t + k_o^2 \bar{\bar{\varepsilon}}_{rt} H_t = -\bar{\bar{\varepsilon}}_{rt}^{-1} \nabla_t \bar{\bar{\varepsilon}}_{rt} \times (\nabla_t \times H_t) \tag{4.8}$$

Here, the subscript "t" refers to the transverse components and $\bar{\bar{\varepsilon}}_{rt}$ is the transverse component of the relative dielectric tensor and can be defined as follows:

$$\bar{\bar{\varepsilon}}_{rt} = \begin{pmatrix} \varepsilon_{xx} & \varepsilon_{xy} \\ \varepsilon_{yx} & \varepsilon_{yy} \end{pmatrix} \tag{4.9}$$

Assuming a z dependence of $e^{-j\beta z}$ for all fields and applying the divergence relation $\nabla \cdot H = 0$, the longitudinal component H_z can be computed from the transverse component H_x and H_y as follows:

$$H_z = \frac{1}{j\beta}\left(\frac{\partial H_x}{\partial x} + \frac{\partial H_y}{\partial y}\right). \tag{4.10}$$

Here, β is the propagation constant of the propagated mode. Using the vector wave equation (4.8) and (4.10) and the divergence relation $\nabla \cdot H = 0$, one can obtain, after some algebraic treatment, the following full-vector eigenvalue equation [1]:

$$\begin{bmatrix} A_{xx} & A_{xy} \\ A_{yx} & A_{yy} \end{bmatrix}\begin{bmatrix} H_x \\ H_y \end{bmatrix} = \beta^2 \begin{bmatrix} H_x \\ H_y \end{bmatrix}. \tag{4.11}$$

Here, A_{xx}, A_{xy}, A_{yx}, and A_{yy} are the differential operators that can be defined such that [1]

$$A_{xx}H_x = \frac{\partial^2 H_x}{\partial x^2} + \frac{\varepsilon_{yy}}{\varepsilon_{zz}}\frac{\partial^2 H_x}{\partial y^2} + \frac{\varepsilon_{yx}}{\varepsilon_{zz}}\frac{\partial^2 H_x}{\partial y \partial x} + k^2 \varepsilon_{yy} H_x. \tag{4.12}$$

$$A_{xy}H_y = \left(1 - \frac{\varepsilon_{yy}}{\varepsilon_{zz}}\right)\frac{\partial^2 H_y}{\partial x \partial y} - \frac{\varepsilon_{yx}}{\varepsilon_{zz}}\frac{\partial^2 H_y}{\partial x^2} - k^2 \varepsilon_{yx} H_y. \tag{4.13}$$

$$A_{yx}H_x = \left(1 - \frac{\varepsilon_{xx}}{\varepsilon_{zz}}\right)\frac{\partial^2 H_x}{\partial y \partial x} - \frac{\varepsilon_{xy}}{\varepsilon_{zz}}\frac{\partial^2 H_x}{\partial y^2} - k^2 \varepsilon_{xy} H_x. \tag{4.14}$$

$$A_{yy}H_y = \frac{\partial^2 H_y}{\partial y^2} + \frac{\varepsilon_{xx}}{\varepsilon_{zz}}\frac{\partial^2 H_y}{\partial x^2} + \frac{\varepsilon_{xy}}{\varepsilon_{zz}}\frac{\partial^2 H_y}{\partial x \partial y} + k^2 \varepsilon_{xx} H_y. \tag{4.15}$$

All the differential operators used in the above equations can be approximated by using the FDMs [1]. Equation (4.11) is a full-vector eigenvalue equation, which describes the modes of propagation for anisotropic optical waveguides. The two transverse field components H_x and H_y are the eigenvectors, and the corresponding eigenvalue is β^2.

Once all three components of the magnetic field H are known, the electric flux density D can be found by applying $\nabla \times H = j\omega D$, which gives the following:

$$D_x = -\frac{1}{\omega\beta}\left(\frac{\partial^2 H_x}{\partial y \partial x} + \frac{\partial^2 H_y}{\partial y^2}\right) + \frac{\beta}{\omega} H_y. \tag{4.16}$$

$$D_y = +\frac{1}{\omega\beta}\left(\frac{\partial^2 H_y}{\partial x \partial y} + \frac{\partial^2 H_x}{\partial x^2}\right) - \frac{\beta}{\omega} H_x. \tag{4.17}$$

$$D_z = \frac{j}{\omega}\left(\frac{\partial H_x}{\partial y} - \frac{\partial H_y}{\partial x}\right). \tag{4.18}$$

The electric field can be computed from D by using $E = \varepsilon^{-1} D$ as follows:

$$\begin{bmatrix} E_x \\ E_y \end{bmatrix} = \frac{1}{\varepsilon_o \left(\varepsilon_{xx} \varepsilon_{yy} - \varepsilon_{xy} \varepsilon_{yx} \right)} \begin{bmatrix} \varepsilon_{yy} & -\varepsilon_{xy} \\ -\varepsilon_{yx} & \varepsilon_{xx} \end{bmatrix} \begin{bmatrix} D_x \\ D_y \end{bmatrix}. \tag{4.19}$$

$$E_z = \frac{1}{\varepsilon_o \varepsilon_{zz}} D_z = \frac{j}{\omega \varepsilon_o \varepsilon_{zz}} \left(\frac{\partial H_x}{\partial y} - \frac{\partial H_y}{\partial x} \right). \tag{4.20}$$

4.3.3 Boundary Conditions

To simulate real structures and create a practical solver, the effects of the simulation boundaries should be fully considered. The basic BPM and FDM boundary conditions set the field values just outside the simulation area to zero, simulating a perfectly conducting metal box. However, the perfect matched layer (PML) [6] is an artificial absorbing layer and is commonly used to simulate problems with open boundaries, especially in the FDM and FEM methods. The key feature of a PML that differentiates it from any other common absorbing material is that it is designed so that the propagating waves from a non-PML medium toward the PML medium do not reflect at the interface. This characteristic permits the PML to effectively absorb the waves leaving the interior of a computational region without reflecting them back into the interior.

The original formulation of PML was investigated by Berenger [6] with the aid of Maxwell's equations and, since that time, there have been several related reformulations of PML for both Maxwell's equations and other wave equations. Berenger's original formulation is called a split-field PML, because it is based on splitting the electromagnetic fields into two unphysical fields in the PML region. Later, a simple and efficient formulation is presented in Ref. [7] and is called the uniaxial PML (UPML) [7]. This formalism is based on representing the PML medium as an artificial anisotropic absorbing material. At first, both Berenger's formulation and the UPML formulation were deduced by manually finding the conditions under which incident plane waves from a homogeneous medium do not reflect from the PML interface. However, both formulations were later shown to be equivalent to a much more wonderful and comprehensive approach: the stretched-coordinate PML [8, 9]. Particularly, PMLs exhibit coordinate transformations in which one (or more) coordinates are transformed to complex numbers. More technically, the analytical solution of the wave equation into complex coordinates was realized by replacing the propagating waves by exponentially decaying waves. This technique permits the formulation of PMLs for inhomogeneous media, other coordinate systems and other wave equations.

As an example for the transformation of PML in the stretched coordinate system. Wherever, the derivative $\partial / \partial x$ appears in the wave equation, it is substituted by [9]

$$\frac{\partial}{\partial x} = \frac{1}{1 + \dfrac{i\sigma(x)}{\omega}} \frac{\partial}{\partial x}, \tag{4.21}$$

where ω is the angular frequency and σ is some function of x. This can be attained by stratifying the following transformation to complex coordinates: $x \rightarrow x + \dfrac{i}{\omega} \int \sigma(x) dx$, or equivalently $dx \to dx \left(1 + \dfrac{i\sigma(x)}{\omega} \right)$. Therefore, whenever σ is positive propagating waves are attenuated because

$$e^{i(kx-\omega t)} \to e^{i(kx-\omega t)-\frac{k}{\omega}\int \sigma(x)dx}, \tag{4.22}$$

where a plane wave propagating in the $+x$ direction (for $k>0$) has been taken.

4.3.4 Maxwell's Equations in Complex Coordinate

Assuming the following change of variables [9],

$$\tilde{x} = \int_0^x S_x(x')dx' \tag{4.23a}$$

$$\tilde{y} = \int_0^y S_y(y')dy' \tag{4.23b}$$

$$\tilde{z} = \int_0^z S_z(z')dz'. \tag{4.23c}$$

Then, the ∇ operator turns into

$$\tilde{\nabla} = \hat{x}\frac{\partial}{\partial \tilde{x}} + \hat{y}\frac{\partial}{\partial \tilde{y}} + \hat{z}\frac{\partial}{\partial \tilde{z}}, \tag{4.24}$$

where

$$\frac{\partial}{\partial \tilde{x}} = \frac{1}{S_x}\frac{\partial}{\partial x}, \frac{\partial}{\partial \tilde{y}} = \frac{1}{S_y}\frac{\partial}{\partial y}, \frac{\partial}{\partial \tilde{z}} = \frac{1}{S_z}\frac{\partial}{\partial z}. \tag{4.25}$$

Under the change of variables in Eq. (4.23) Maxwell's equations (4.1) to (2.4) can be reformulated as follows [9]:

$$\tilde{\nabla} \times E = -j\omega\mu H \tag{4.26}$$

$$\tilde{\nabla} \times H = j\omega\bar{\bar{\varepsilon}} E \tag{4.27}$$

$$\tilde{\nabla} \cdot \left(\bar{\bar{\varepsilon}} E \right) = 0 \tag{4.28}$$

$$\tilde{\nabla} \cdot \mu H = 0. \tag{4.29}$$

And $\tilde{x}, \tilde{y},$ and $\tilde{z}$ can have complex values. The most motivating case occurs when $\tilde{x}, \tilde{y},$ and $\tilde{z}$ are complex. In this case, the waves traveling through the complex coordinates are weakened, but not reflected.

(a)   (b)

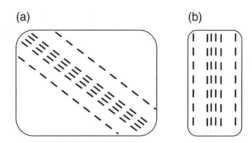

Figure 4.2 (a) Typical distribution of the nonzero matrix elements in the finite-difference method and (b) more economical band storage

4.3.5 Matrix Solution

The operators A_{xx}, A_{xy}, A_{yx}, and A_{yy} in Eq. (4.11) can be displaced by a $N \times N$ sparse matrix, where $N = (n_x + 1)(n_y + 1)$ and n_x and n_y are the number of mesh points in the x and y directions, respectively. When these four sparse matrices are gathered in matrix A as in Eq. (4.11), the modes and propagation constants of the waveguide can be calculated from the eigenvectors and eigenvalues of the consequent matrix. The resultant matrix A is very sparse because each difference equation involves the field at only a very few mesh points. Figure 4.2 shows a typical distribution for the nonzero elements of A [10, 11]. In the next subsection, the power methods will be explained in detail.

It should be noted that the full-vector eigenvalue equation (4.11) is solved numerically for a set of modes, and the dominant mode is defined as the mode with the highest real effective index value. Since, $\nabla \cdot H = 0$ and interface boundary conditions are automatically satisfied in the formulation, then there is no chance for spurious (nonphysical) modes to appear in the spectrum of the solution. In addition, the PML boundary condition is employed at the edges of the computational window in order to account for the leakage property of the modes. For this objective, the spacing between the meshing points at the boundaries is complex to incorporate perfectly matched layers using the complex coordinate stretching method [8, 9].

4.3.5.1 Power Method

The power method is applied to calculate the eigenvalue of largest magnitude of a given matrix [12]. Equation (4.11) can be reformulated as $AX = v\,X$ and applied to a matrix A, the power method is based on choosing a vector X and forming the sequence $c_0 X$, $c_1 AX$, $c_2 A^2 X$, $c_3 A^3 X$, ... where c_0, c_1, c_2, ..., c_N, are the scaling constants chosen to dodge computer surplus due to much large vector components. The sequence will generally converge to an eigenvector of A, and the eigenvalue will be clear too with smart choice of scaling constants. This eigenvalue is usually the dominant eigenvalue of A, the one having the greatest absolute value provided such an eigenvalue is real. The usual operation of the power method can be described as follows:

1. Set X_0 so that its largest component in absolute value is 1, and set $v_0 = 0$ as the first approximation to the eigenvalue. Also, determine the required precision (PRE) for the eigenvalue, and the required maximum number of iterations and set an iteration counter $k = 1$.

2. Compute $Y_k = AX_{k-1}$ and find the largest component in absolute value of Y_k. Denote it as v_k.
3. Fix $X_k = (1/v_k)\,Y_k$ and if $|v_k - v_{k-1}| < \text{PRE}$, the iteration will be terminated and the eigenvalue and associated eigenvector are v_k and X_k. Otherwise, go on to the next step.
4. Boost k by 1. If k is greater than the maximum number of iterations to be performed, stop. Otherwise, return to the second step.

4.3.5.2   Inverse Power Method

This method [12] can be applied to investigate the smallest (magnitude) eigenvalue v_N by stratifying the power method to the inverse matrix A^{-1} provided the matrix is non-singular and taking the inverse of the largest component of the limit. The basic idea is to rewrite the equation $AX = v\,X$ as $A^{-1}X = v^{-1}X$. In this way, the procedure will converge to the dominant eigenvalue of A^{-1}, the reciprocal of which is the eigenvalue of A having the smallest absolute value. The operation steps of this method are identical to those of the power method with the exception of the following:

In the second step: the equation, $Y_k = A^{-1}X_{k-1}$ is calculated by solving the system $AY_k = X_{k-1}$, using LU decomposition. If this system does not have a unique solution, the iteration is terminated and the eigenvalue of A is zero.

4.3.5.3   Shifted Inverse Power Method

The objective of the shifted inverse power method is to find the eigenvalue that is not necessarily of the largest or smallest magnitude. In particular, it is used to find all real eigenvalues of a matrix if estimates of their locations are available. If the power method is applied to $(A - uI)^{-1}$ where u is an estimate of the eigenvalue v, the resulting eigenvalue and the eigenvector will be v' and X, respectively. Then, the eigenvalue and eigenvector of matrix A will be $(1/v') + u$ and X, respectively.

4.4   Summary

The chapter starts with a brief overview of the existing modeling methods followed by the formulation of the FDM [1] with perfect boundary conditions capabilities which is employed in different studies throughout this book.

References

[1] Fallahkhair, A.B., Li, K.S. and Murphy, T.E. (2008) Vector finite difference modesolver for anisotropic dielectric waveguides. *J. Lightwave Technol.*, **26** (11), 1423–1431.
[2] Obayya, S.S.A., Rahman, B.M.A. and Grattan, K.T.V. (2005) Accurate finite element modal solution of photonic crystal fibres. *IEE Proc. Optoelectron.*, **152**, 241–246.
[3] Campbell, S., McPhedran, R.C., Martijn de Sterke, C. and Botten, L.C. (2004) Differential multipole method for microstructured optical fibers. *J. Opt. Soc. Am. B*, **21** (21), 1919–1928.
[4] De Francisco, C.A., Borges, B.V. and Romero, M.A. (2003) A semivectorial method for the modeling of photonic crystal fibers. *Microw. Opt. Technol. Lett.*, **38** (5), 418–421.

[5] Obayya, S.S.A., Azizur Rahman, B.M., Grattan, K.T.V. and El-Mikati, H.A. (2002) Full vectorial finite-element-based imaginary distance beam propagation solution of complex modes in optical waveguides. *J. Lightwave Technol.*, **20**, 1054.

[6] Berenger, J. (1994) A perfectly matched layer for the absorption of electromagnetic waves. *J. Comput. Phys.*, **114**, 185–200.

[7] Gedney, S.D. (1996) An anisotropic perfectly matched layer absorbing media for the truncation of FDTD latices. *IEEE Trans. Antennas Propag.*, **44**, 1630–1639.

[8] Chew, W.C. and Weedon, W.H. (1994) A 3d perfectly matched medium from modified Maxwell's equations with stretched coordinates. *Microw. Opt. Technol. Lett.*, **7**, 590–604.

[9] Chew, W.C., Jin, J.M. and Michielssen, E. (1997) Complex coordinate stretching as a generalized absorbing boundary condition. *Microw. Opt. Technol. Lett.*, **15** (6), 363–369.

[10] Lehoucq, R.B., Sorensen, D.C. and Yang, C. (1998) *ARPACK Users' Guide: Solution of Large-Scale Eigenvalue Problems with Implicitly Restarted ArnoldiMethodss*, SIAM, Philadelphia.

[11] Davis, T.A. and Duff, I.S. (1997) An unsymmetric-pattern multifrontal method for sparse LU factorization. *SIAM J. Matrix Anal. Appl.*, **18** (1), 140–158.

[12] Bronson, R. (1989) *Schaum's Outline of Theory and Problems of Matrix Operations*, McGraw-Hill, New York.

5

Assessment of the Full-Vectorial Finite-Difference Method

5.1 Introduction

In this chapter, the numerical precision of the full-vectorial finite-difference method (FVFDM) [1] is demonstrated through the modal analysis of a soft glass photonic crystal fiber (PCF) infiltrated with nematic liquid crystal (NLC) material [2–4]. An overview of the previously reported LC-PCF and background of the soft glasses are first given. Then, the different modal properties of the suggested design, such as effective index, birefringence, effective mode area, hybridness, and nonlinearity are explained in detail. Finally, the fabrication aspects of the proposed design will be described.

5.2 Overview of the LC-PCF

The wave guiding through bulk LC is limited due to its high attenuation in the range 20–40 dB/cm [5]. Therefore, PCF infiltrated with LC is attracting the interest of many researchers all over the world. In this case, the light will be propagated through the core region with small attenuation. A PCF structure infiltrated with LC has unique propagation and polarization properties. LC-PCFs for single polarization or high-birefringence guidance can be achieved [6]. Switching between index- and photonic bandgap (PBG)-guiding mechanisms can occur in the same structure by the control of the temperature and an external electric field [7, 8]. Moreover, the PBG of the crystal can be modified by the tunability of the optical properties of the LC with temperature or an external electric field [9]. Electrical control can be achieved using externally applied electric fields to reorient the LC molecules [8, 10]. The LC-PCF can be placed between electrodes and the PBGs can be electrically tuned [10]. The reorientation of the LC by the electric filed offers high anisotropy in the cross section of the PCF with polarization-dependent loss (PDL). However, the LC-PCF has no PDL when the E-field is off.

Computational Liquid Crystal Photonics: Fundamentals, Modelling and Applications, First Edition.
Salah Obayya, Mohamed Farhat O. Hameed and Nihal F.F. Areed.
© 2016 John Wiley & Sons, Ltd. Published 2016 by John Wiley & Sons, Ltd.

The propagation properties of polymer PCF infiltrated with LC are presented by Woliñski *et al.* [11]. It has been reported that with increasing temperature, the PBGs in the transmission spectrum shift toward short wavelengths and become narrower. As the temperature approaches the nematic–isotropic phase transition, the PBGs shift back toward long wavelengths and become wider.

The unique characteristics of the LC-PCF with large LC electro-optic effect can be used for different applications, such as switches [9], tunable wave plates [12], and polarization rotators [13, 14]. Additionally, an all-optical modulator has been demonstrated by Alkeskjold *et al.* [15], which utilizes a pulsed 532 nm laser to modulate the spectral position of the bandgaps in a PCF filled with a dye-doped NLC; an LC-PBG fiber-based broadband polarimeter at the operating wavelength of 1550 nm has also been introduced in Ref. [16]; and a compact electrically controlled broadband LC-PBG fiber polarizer has been designed and fabricated by Wei *et al.* [17].

In this chapter, the FVFDM [1] is used to study and analyze a novel design of highly nonlinear index-guiding soft glass PCF with a liquid crystal core (SGLC-PCF) [3, 4]. The suggested design depends on using soft glass and NLC of types SF57 (lead silica) and E7, respectively. The refractive index of the SF57 material is greater than the ordinary n_o and extraordinary n_e refractive indices of the E7 material, which guarantees the index guiding of the light through the suggested SGLC-PCF. For such a structure, the effective index, birefringence, effective mode area, and nonlinearity of the two fundamental polarized modes have been studied thoroughly. The effect of structure geometrical parameters, the rotation angle of the director of the NLC and temperature on the different modal properties is investigated. In addition, the birefringence is tailored so that high birefringence of 0.042 at the operating wavelength $\lambda = 1.55$ μm can be achieved. Moreover, the reported design also offers high nonlinearities of 425.5 and 470.8 W^{-1} km^{-1} for the quasi-transverse electric (TE) and quasi-transverse magnetic (TM) modes, respectively, at a wavelength of 1.55 μm.

5.3   Soft Glass

The novel optical properties of PCFs can result from the new combination of the wavelength-scaled features of the microstructured cladding and the large index contrast between the core and the cladding region. The non-silica glasses, such as soft glass [18–20] and tellurite glass [21], have different and unique material properties that can be used for PCF fabrication. For example, the soft glass has a higher refractive index and nonlinearity than silica glass. In addition, soft glass has high IR transmittance and large rare-earth solubility. The novel combination of non-silica highly nonlinear glasses with a tight core results in a further dramatic increase in the fiber nonlinearity. Therefore, non-silica PCF can be used for highly nonlinear applications. Moreover, PCFs based on non-silica glasses have good transparency in the infrared (IR) with greater spectral range. Thus, non-silica glasses can be used to obtain practical fibers in various wavelength ranges [18], especially the mid-IR region for soft glasses.

PCFs that are made of soft glass [18, 19] have widespread interest since they offer optical properties that cannot be provided by silica, such as high nonlinearity, high refractive index, high rare-earth solubility, and mid-IR transmission. The soft glass of type SF57 [22] is a commercial lead silicate glass that has the highest nonlinearity among the commercially available lead silicate glasses. Additionally, it provides easy fiber fabrication due to its good mechanical, chemical, and thermal stability. It has a low processing temperature of approximately 5200°C; therefore, the PCF preform can be extruded directly from the bulk glass. It should be noted that the softening temperature for silica glass is 1,500–16,000°C. Soft glass PCFs based on lead silicate glasses [18, 23], heavy metal oxide glasses [21], and lanthanum–gallium–sulfide glass [24] have been fabricated successfully. The techniques used include extrusion, capillary stacking, ultrasonic drilling, and build-in casting. However, the extrusion technique can fabricate complex PCF structures with a good surface quality that cannot be produced using capillary stacking . The extrusion approach has been used to fabricate PCF based on polymer material [25], tellurite [21, 26], lead silicate [23, 27, 28], bismuth silicate [29], and chalcogenide glasses [24].

5.4   Design of Soft Glass PCF with LC Core

Figure 5.1 shows a cross section of the suggested triangular lattice SGLC-PCF with air holes in the cladding region [3]. All the cladding holes have the same diameter d and are arranged with a hole pitch Λ. The central hole of diameter d_o is infiltrated with an NLC of type E7. The reported structure [3] is different from the NLC-PCF [2] in which all the cladding holes have been infiltrated by the NLC with a solid soft glass core. Therefore, the new design has the advantage of ease of fabrication. The NLCs used in the proposed structure are anisotropic materials consisting of rod-like molecules, which are characterized by ordinary index n_o and extraordinary index n_e. The ordinary n_o and extraordinary n_e refractive indices of the E7 material were measured previously by Li $et\ al.$ [30] at different visible wavelengths in the temperature range from 15 to 50°C with a step of 5°C. Then, the Cauchy model was used to fit the measured n_o and n_e, which can be described as follows [30]:

$$n_e = A_e + \frac{B_e}{\lambda^2} + \frac{C_e}{\lambda^4}. \tag{5.1}$$

$$n_o = A_o + \frac{B_o}{\lambda^2} + \frac{C_o}{\lambda^4}. \tag{5.2}$$

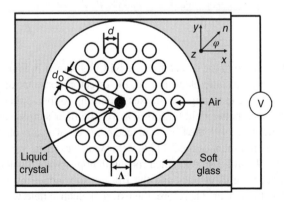

Figure 5.1   Cross section of the SGLC-PCF sandwiched between two electrodes. *Source*: Ref. [3]

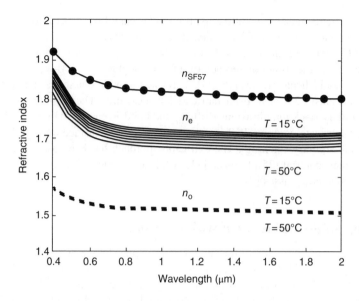

Figure 5.2 Variation of n_o and n_e of the E7 material with the wavelength at different temperatures T from 15 to 50°C with a step of 5°C. The solid line with closed circles represents the variation of the refractive index of the SF57 material n_{SF57} with the wavelength. *Source*: Ref. [2]

Here, A_e, B_e, C_e, A_o, B_o, and C_o are the coefficients of the Cauchy model. The Cauchy coefficients at $T = 25°C$ are $A_e = 1.6933$, $B_e = 0.0078$ μm², $C_e = 0.0028$ μm⁴, $A_o = 1.4994$, $B_o = 0.0070$ μm², and $C_o = 0.0004$ μm⁴. The variations in n_o and n_e of the E7 material with the wavelength at different temperatures T from 15 to 50°C with a step of 5°C are shown in Figure 5.2. It is evident from the figure, that n_e is greater than n_o at the measured temperatures within the reported wavelength range. In the proposed design, the relative permittivity tensor ε_r of the E7 material is taken as [31]

$$\varepsilon_r = \begin{pmatrix} n_o^2 \sin^2 \varphi + n_e^2 \cos^2 \varphi & \left(n_e^2 - n_o^2\right)\cos\varphi\sin\varphi & 0 \\ \left(n_e^2 - n_o^2\right)\cos\varphi\sin\varphi & n_o^2 \cos^2 \varphi + n_e^2 \sin^2 \varphi & 0 \\ 0 & 0 & n_o^2 \end{pmatrix}, \tag{5.3}$$

where φ is the rotation angle of the director of the NLC, as shown in Figure 5.1.

Under the application of a static electric field, the director's orientation can be controlled, since the NLC molecules tend to align their axes according to the applied field. This can be achieved successfully with better field uniformity over the fiber's cross section, as described by Haakestad *et al.* [9]. In Ref. [9], the fiber is placed between two pairs of electrodes allowing arbitrary control of the alignment of the NLC director via an external voltage, as shown schematically in Figure 5.1. In addition, two silica rods with appropriate diameter are used to control the spacing between the electrodes and the fiber is surrounded by silicone oil that has higher dielectric strength than air. Therefore, the external electric field will be uniform across the fiber cross section, which results in good alignment of the director of the NLC. Further, the nonuniform electric field region will be at the edges away from the core regions, and so will have little effect on the performance of the proposed fiber. Other layouts, such as those described in Refs. [8, 32] can also be used to ensure better field uniformity over the fiber's cross section.

The proposed in-plane alignment of the NLC can be exhibited under the influence of appropriate homeotropic anchoring conditions [6, 31]. Haakestad *et al.* [9] demonstrated experimentally that in the strong field limit, the NLC of type E7 is aligned in plane in capillaries of diameter 5 μm. Additionally, Alkeskjold and Bjarklev [16] demonstrated the in-plane alignment of the E7 material in PCF capillaries of diameter 3 μm with three different rotation angles, 0°, 45°, and 90° using two sets of electrodes.

The background of the reported SGLC-PCF [3] is a soft glass of type SF57 (lead silica). The wavelength-dependent refractive index of the SF57 material is also shown in Figure 5.2. It can be seen that the refractive index of the SF57 material is greater than the n_o and n_e of the E7 material. Therefore, the propagation occurs through the core region by modified total internal reflection. The Sellmeier equation for the soft glass of type SF57 [28] is given by

$$n_{SF57}^2 = A_o + A_1\lambda^2 + \frac{A_2}{\lambda^2} + \frac{A_3}{\lambda^4} + \frac{A_4}{\lambda^6} + \frac{A_5}{\lambda^8}, \tag{5.4}$$

where n_{SF57} is the refractive index of the SF57 material, $A_o = 3.24748$, $A_1 = -0.0096$ μm^{-2}, $A_2 = 0.0494$ μm^2, $A_3 = 0.00294$ μm^4, $A_4 = -1.4814 \times 10^{-4}$ μm^6, and $A_5 = 2.7843 \times 10^{-5}$ μm^8 [28].

5.5 Numerical Results

5.5.1 FVFDM Validation

To validate the simulation results, initially the conventional triangular lattice PCF with air holes and silica background reported in Ref. [33] was considered. In this study, the hole pitch and hole radius are fixed to 2.3 and 0.575 μm, respectively. In addition, the refractive index of the silica was taken as 1.45 at $\lambda = 1.55$ μm. Figure 5.3 shows the wavelength dependence of the real part of the complex effective index of the quasi-TE mode of the conventional PCF. As can be seen from the figure, there is an excellent agreement between the FVFDM results and those reported in the literature [33]. Next, the NLC-PCF [2] was analyzed by using the full vector *H*-field finite-element mode solver based on the variational technique (finite element method or FEM) [34] as well as the FVFDM. In this study, all the holes of the NLC-PCF have the same radius r and are arranged with a hole pitch Λ of 2.3 μm. In addition, n_o, n_e, and n_{SF57} are taken as 1.5024, 1.6970, and 1.802, respectively, at $\lambda = 1.55$ μm. Moreover, the rotation angle and the temperature are fixed to 90° and 25°C, respectively. The variation of the real part of the complex effective index of the quasi-TM mode with the wavelength at different r, 0.6, 0.7, 0.8, and 0.9 μm is also shown in Figure 5.3. It is evident from the figure that there is excellent agreement between the results of the FVFDM and those obtained by the FEM [34] at $r = 0.6$ and 0.7 μm. It is can also be seen from Figure 5.3 that the effective index of the quasi-TM mode decreases with increasing wavelength due to less confinement of the mode through the core region at long wavelength. Moreover, the effective refractive index of the cladding

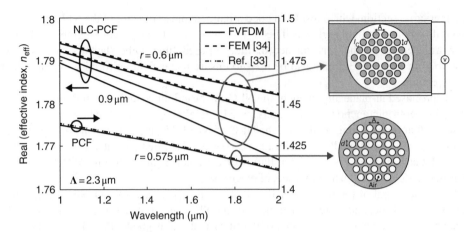

Figure 5.3 Variation of the real part of the complex effective index of the quasi-TM modes of the NLC-PCF (*Source*: Ref. [2]) with the wavelength at different cladding hole radius *r*. The variation of the real part of the complex effective index of the quasi-TE mode of the conventional silica PCF (*Source*: Ref. [33]) with the wavelength is also shown in the figure

region, and hence the effective index of the quasi-TM mode decreases with increase in the infiltrated NLC hole radius.

5.5.2   Modal Hybridness

In the proposed structure shown in Figure 5.1 [3], all the cladding air holes have the same diameter d and are arranged with a hole pitch $\Lambda = 5$ μm and d/Λ ratio of 0.6, while the central hole diameter d_o is taken as 1.0 μm. In addition, n_o and n_e of the E7 material are fixed to 1.5024 and 1.6970, respectively, at the operating wavelength $\lambda = 1.55$ μm and at a temperature of 25°C. The rotation angle of the director of the NLC is taken as 45° and n_{SF57} is fixed to 1.802 at $\lambda = 1.55$ μm. To understand the effect of the infiltration of the NLC, a soft glass PCF with central air hole is first considered. Figure 5.4a and b shows the dominant H_y and nondominant H_x field components of the quasi-TE mode. Simply, the fundamental quasi-TE mode refers to the fundamental H_{11}^y or E_{11}^x modes, while the fundamental quasi-TM mode refers to the fundamental H_{11}^x or E_{11}^y modes according to the Cartesian coordinates shown in Figure 5.1. It is observed from Figure 5.4a and b that the dominant H_y field profile is symmetric, while the nondominant H_x field component is clearly shown to be antisymmetric and its maximum magnitude is only 0.03, normalized to the maximum value of H_y. Next, the SGLC-PCF with NLC core is considered. The dominant H_y and nondominant H_x field profiles of the quasi-TE mode are shown in Figure 5.4c and d, respectively. It can be noted that the field profiles of the dominant and non-dominant components of the quasi-TE mode are very similar. The hybridness is defined as the nondominant to dominant field ratios, that is, hybridness = (h_x/h_y) for the quasi-TE mode, while it is equal to (h_y/h_x) for the quasi-TM mode, where h_x and h_y are the maximum amplitudes of the field components. The maximum value of H_x is 0.99 normalized to the maximum value of the dominant H_y component. This means that the proposed SGLC-PCF with the NLC supports highly hybrid modes that can be used in designing polarization conversion devices.

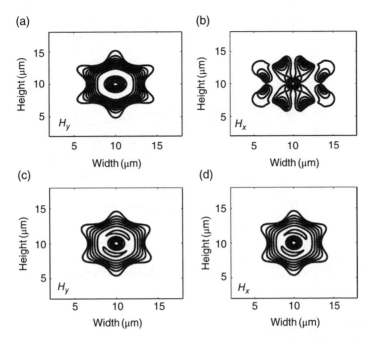

Figure 5.4 Contour plot of the dominant H_y and nondominant H_x field profiles of the fundamental quasi-TE mode for (a and b) the soft glass PCF with central air hole and for (c and d) the SGLC-PCF. *Source*: Ref. [3]

5.5.3 Effective Index

The influences of the structure geometrical parameters, temperature, and rotation angle φ of the director of the NLC on the modal properties of the SGLC-PCF are investigated in detail. The analyzed parameters are the effective index, birefringence, effective mode area, and non-linearity. Figure 5.5 shows the wavelength dependence of the real part of the complex effective indices of the quasi-TE and quasi-TM modes of the SGLC-PCF at different cladding hole radii r, 0.7, 0.8, and 0.9 μm. In this study, the hole pitch Λ, central hole radius r_o, rotation angle, and temperature are taken as 2.3 μm, 0.5 μm, 90°, and 25°C, respectively. In addition, n_o, n_e, and n_{SF57} are fixed to 1.5024, 1.6970, and 1.802, respectively, at $\lambda = 1.55$ μm. It can be seen from Figure 5.5 that the effective indices of the two polarized modes decrease with increasing wavelength due to less confinement of the modes through the core region at long wavelength. Further, the effective refractive index of the cladding region decreases with increasing air hole radius. Therefore, the effective indices of the two polarized modes decrease with increasing cladding air hole radius as can be seen in Figure 5.5. It is also evident from Figure 5.5 that the effective index of the quasi-TM mode is greater than that of the quasi-TE mode. At $\varphi = 90°$, the permittivity tensor ε_r of the E7 material is the diagonal of $[\varepsilon_{xx}, \varepsilon_{yy}, \varepsilon_{zz}]$, where $\varepsilon_{xx} = n_o^2$, $\varepsilon_{yy} = n_e^2$, and $\varepsilon_{zz} = n_o^2$. In this case, ε_{xx} is less than ε_{yy}; therefore, the effective refractive index of the core region seen by the quasi-TM mode is greater than that seen by the quasi-TE mode. Consequently, the effective index of the quasi-TM modes is greater than that of the quasi-TE mode.

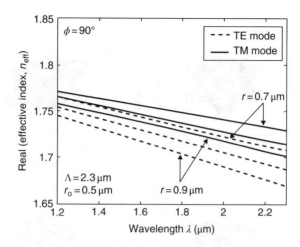

Figure 5.5 Variation of the real part of the complex effective indices of the quasi-TE and quasi-TM modes of the SGLC-PCF with the wavelength at different cladding air hole radii r, 0.7, 0.8, and 0.9 μm. *Source*: Ref. [3]

5.5.4 Effective Mode Area

The effective mode area [35, 36] can be defined as a quantitative measure of the area that a waveguide or fiber mode effectively covers in the transverse dimensions. This spatial extension of the mode is of great importance to several applications. For nonlinear applications, one would like to have a very confined mode, whereas for high power applications one would typically prefer a large mode to avoid nonlinear effects and material damage. The effective mode area A_{eff} of the PCFs can be calculated using [35, 36]

$$A_{\text{eff}} = \frac{\left(\iint\limits_{S} |H_t|^2 \, dxdy\right)^2}{\iint\limits_{S} |H_t|^4 \, dxdy},$$

(5.5)

where H_t is the transverse magnetic field vector and S is the area enclosed within the computational window.

5.5.5 Nonlinearity

Nonlinear effects in fiber can be used for a wide range of optical processing applications in telecommunications, such as optical data regeneration, optical demultiplexing, and wavelength conversion. Consequently, there is a great interest in developing fibers with high effective nonlinearity per unit length in order to reduce device lengths and the associated optical power requirements for fiber-based nonlinear devices. One of the most exciting prospects for PCF technology is the development of fibers with large values of optical nonlinearity per unit length γ. The effective nonlinearity γ [37] is commonly used to measure the fiber nonlinearity and is given by

$$\gamma = \frac{2\pi n_2}{A_{eff} \lambda}, \tag{5.6}$$

where n_2 is the nonlinear coefficient for the material, A_{eff} is the effective mode area, and λ is the optical wavelength. Fibers with small core dimension and cladding with a large air fill fraction allow extremely tight mode confinement, that is, a small effective area, and hence a high value of γ. For example, standard single-mode fiber, SMF28 fiber has an A_{eff} of approximately 90 μm^2 at a wavelength of 1550 nm, and since n_2 of silica is approximately 2.2×10^{-20} m^2/W, γ is of the order of 1 W^{-1} km^{-1} [37]. In conventional step-index fibers, the nonlinearity has been increased to 20 W^{-1} km^{-1} by decreasing the core size and increasing the germanium concentration in the core region [38]. In small-core PCFs, tight mode confinement is provided by the large index contrast between air and glass. The pure silica PCFs that equate $A_{eff} \sim 1.3$ μm^2 with values of $\gamma = 70$ W^{-1} km^{-1} at a wavelength of 1550 nm, which is approximately 3.5 times greater than the highest reported nonlinearity in conventional fibers, have been demonstrated in Refs. [39, 40].

The nonlinearity that can be achieved in any fiber is limited by its mode confinement. Significantly higher values of γ can be achieved by combining tight mode confinement with the use of non-silica glasses with greater intrinsic material nonlinearity coefficients than silica. In order to achieve higher values of γ, materials with larger values of n_2 are particularly promising candidates. The n_2 of soft glasses can be more than an order of magnitude larger than that of silica. Examples of suitable glasses that have been used to make PCFs include chalcogenide [24], tellurite [21], bismuth oxide [41], and lead silicate glasses [42]. It has been shown that it is possible to extend PCF fabrication techniques to such glasses and the SF57 lead silicate holey fibers (HFs) with a record nonlinearity of $\gamma = 1860$ W^{-1} km^{-1} [27], which is more than 1800 times larger than the standard silica SMF.

5.5.6 Birefringence

The birefringence has also been studied for the proposed design. The birefringence can be defined as the difference between the real part of the effective indices n_{effTE} and n_{effTM} of the quasi-TE and quasi-TM modes, respectively. Figure 5.6 shows the variation of the effective mode area of the quasi-TE modes and birefringence with the wavelength at different cladding hole radii r, 0.7, 0.8, and 0.9 μm. It can be seen from the figure that the birefringence increases with increasing cladding hole radius. The proposed design with $r = 0.9$ μm offers high birefringence of 0.02 at $\lambda = 1.55$ μm. As the cladding hole radius increases, the confinement of the mode through the core region increases. Therefore, A_{eff} of the quasi-TE modes decreases with increasing cladding hole radius, as shown in Figure 5.6. It is also evident from the numerical results that the effective mode area of the quasi-TM mode at $\varphi = 90°$ is less than that of the quasi-TE mode. At $\varphi = 90°$, ε_{xx} is less than ε_{yy}; therefore, the index contrast seen by the quasi-TE modes is less than that seen by the quasi-TM modes. Consequently, the confinement of the quasi-TM modes is more than that of the quasi-TE modes.

The effect of the hole pitch Λ on the modal properties is the next parameter to be considered. The variation of the effective mode area of the quasi-TE mode and birefringence with the wavelength at different hole pitch Λ values, 2.0, 2.1, and 2.2 μm is shown in Figure 5.7. In this study, the cladding hole radius r, NLC infiltrated central hole radius r_o, rotation angle of the

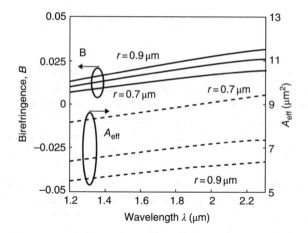

Figure 5.6 Variation of A_{eff} of the quasi-TE mode and birefringence with the wavelength at different cladding hole radii r, 0.7, 0.8, and 0.9 μm. *Source*: Ref. [3]

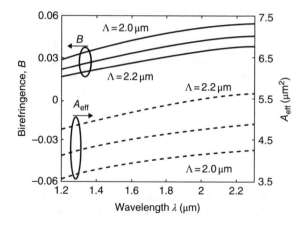

Figure 5.7 Variation of A_{eff} of the quasi-TE mode and birefringence with the wavelength at different hole pitch values Λ, 2.0, 2.1, and 2.2 μm. *Source*: Ref. [3]

NLC, and temperature are taken as 0.9 μm, 0.5 μm, 90°, and 25°C, respectively. As the hole pitch increases, the confinement of the mode through the core region decreases. Therefore, the effective mode area increases with increasing hole pitch, as shown in Figure 5.7. In addition, it is found that the birefringence increases with decreasing hole pitch. The suggested design with $r = 0.9$ μm, $r_o = 0.5$ μm, and $\Lambda = 2.0$ μm offers high birefringence of 0.042 at $\lambda = 1.55$ μm, and this is greater than the high birefringence of 0.0378 of a squeezed hexagonal lattice with elliptical holes [43].

The effect of the NLC-infiltrated central hole radius r_o on the modal properties of the SGLC-PCF is also considered. In this evaluation, the cladding hole radius r, hole pitch, rotation angle, and temperature are fixed to 0.9 μm, 2.0 μm, 90°, and 25°C, respectively. The variation of the nonlinearity of the quasi-TM mode and birefringence with the wavelength at

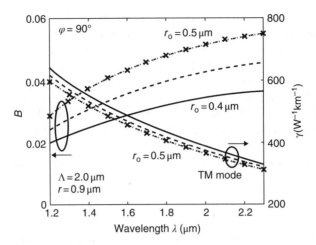

Figure 5.8 Variation of the nonlinearity of the quasi-TM mode and birefringence with the wavelength at different central hole radii r_o, 0.4, 0.45, and 0.5 μm. *Source*: Ref. [3]

different central hole radii r_o, 0.4, 0.45, and 0.5 μm is shown in Figure 5.8. As the central hole radius increases, the index contrast seen by the quasi-TE mode decreases, and hence the confinement of the mode through the core region decreases. Consequently, the effective mode area increases with increasing central hole radius, as shown in Figure 5.8. It is also evident from the figure that the birefringence increases with increasing central hole radius while the nonlinearity decreases.

As r_o increases, the index contrast seen by the quasi-TM mode decreases, and hence the confinement of the mode through the core region decreases. Consequently, the effective mode area increases while the nonlinearity decreases with increasing r_o. The suggested design with $r = 0.9$ μm, $r_o = 0.5$ μm, and $\Lambda = 2.0$ μm offers high birefringence of 0.042 at $\lambda = 1.55$ μm, which is greater than the high birefringence of 0.0378 of a squeezed hexagonal lattice with elliptical holes [43]. In addition, the confinement losses have been found to be 3.2×10^{-11} and 2.8×10^{-12} dB/m at the operating wavelength $\lambda = 1.55$ μm for the quasi-TE and quasi-TM modes, respectively. At $\varphi = 90°$, ε_r is the diagonal of $[n_o^2, n_e^2, n_o^2]$ and ε_{xx} is less than ε_{yy}. Therefore, the index contrast seen by the quasi-TE modes is less than that of the quasi-TM mode. Consequently, the quasi-TM modes are more confined through the core region than the quasi-TE modes. As a result, the confinement losses of the quasi-TM modes are less than those of the quasi-TE modes. The reported design also gives high nonlinearities of 425.5 and 470.8 W^{-1} km^{-1} for the quasi-TE and quasi-TM modes, respectively, at a wavelength of 1.55 μm. However, the solid core soft glass PCF with NLC-filled cladding holes ($r = 0.9$ μm, $\Lambda = 2.0$ μm, and $\varphi = 90°$) [2] gives nonlinearities 506.9 and 348.7 W^{-1} km^{-1} at $\lambda = 1.55$ μm, for the quasi-TE and quasi-TM modes, respectively. However, the reported structure has the advantage of ease of fabrication, as implemented by Xiao *et al.* [44].

It should be noted that the infiltration of the NLC in a hole of diameter 1.0 μm or less (0.7–1.0 μm) and hole pitch of 2.0 μm is achieved experimentally by Woliński *et al.* [45]. In addition, a similar filling technique for only central holes has been suggested in Ref. [46] with dimensions in the same region as the proposed PCF. Moreover, Vieweg *et al.* [47] have recently demonstrated full flexibility of individual closing of holes and subsequent filling of PCFs with highly nonlinear liquids or liquid crystals using a two-photon direct-laser writing technique. Therefore, it is believed that the suggested design can be experimentally fulfilled.

Through the fabrication process, the viscosity of the liquid crystalline state can affect the filling speed of the suggested PCF. In this regard, Scolari [48] reported experiments on the filling length versus filling time for LC of type E7 (filled at the isotropic temperature) infiltrated into PCF. The filling speed was high for the E7 material since its viscosity was lowered by the fact that the LC was filled at its isotropic phase. After 3 h, the filling length was 11.2 cm. Such filling length is enough to be used in communication system devices, such as polarization rotators, polarization splitters, and multiplexers–demultiplexers. In addition, decaying exponential behavior of the filling length as a function of time is observed.

5.5.7   Effect of the NLC Rotation Angle

The rotation angle φ of the director of the NLC can be controlled by an external static electric field; therefore, the effect of the rotation angle on the modal properties is investigated. In this study, the hole pitch, cladding hole radius, central hole radius, and temperature are fixed to 2 μm, 0.9 μm, 0.5 μm, and 25°C, respectively. In this evaluation, the effective indices of the two polarized modes are evaluated at different φ, 0°, 30°, 45°, 60°, and 90°. It is found that the effective indices of the quasi-TE modes are nearly constant in the range of φ from 0° to less than 45°. In addition, the effective indices of the quasi-TE modes are nearly invariant in the range of φ from 45° to 90°. However, the effective indices of the quasi-TE modes in the range of φ from 45° to 90° are less than those in the range from 0° to less than 45°. In the range of φ from 0° to less than 45°, ε_{xx} is more dependent on n_e, while it is more dependent on n_o in the range from 45° to 90°. Therefore, the effective refractive index of the core region, and hence the effective index of the quasi-TE mode in the range of φ from 0° to less than 45°, is greater than those in the range from 45° to 90°. Additionally, the index contrast seen by the quasi-TE mode in the range from 0° to less than 45° is larger than that seen in the range from 45° to 90°. Consequently, the effective mode area in the range from 0° to less than 45° is smaller than that in the range from 45° to 90°. Therefore, the nonlinearity in the range from 0° to less than 45° is larger than that in the range from 45° to 90° as can be seen in Figure 5.9. The figure shows the nonlinearity variation of the quasi-TE mode and birefringence at different φ, 0°, 30°, 60°, and 90° [4]. It is also evident from the figure that the birefringence in the range of φ from 0° to less than 45° is equal to that in the range from 45° to 90° but with opposite sign. At $\varphi = 0°$, the permittivity tensor ε_r of the E7 material is the diagonal of $[n_e^2, n_o^2, n_o^2]$, while at $\varphi = 90°$, ε_r is the diagonal of $[n_o^2, n_e^2, n_o^2]$. It can be noted that ε_{xx} at $\varphi = 0°$ is equal to ε_{yy} at $\varphi = 90°$; therefore, the effective index of the quasi-TE mode at $\varphi = 0°$ is equal to the effective index of the quasi-TM mode at $\varphi = 90°$. In addition, ε_{yy} at $\varphi = 0°$ is equal to ε_{xx} at $\varphi = 90°$; therefore, the effective index of the quasi-TM mode at $\varphi = 0°$ is equal to the effective index of the quasi-TE mode at $\varphi = 90°$. Therefore, the birefringences at $\varphi = 0°$ and 90° are equal but of opposite sign.

The effect of the rotation angle of the director of the NLC on the hybridness is also evaluated, while the hole pitch, cladding hole radius, NLC infiltrated central hole radius, and temperature are taken as 2 μm, 0.9 μm, 0.5 μm, and 25°C, respectively. In addition, the operating wavelength is taken as 1.55 μm. It is found that as φ is increased from 0° to 45°, the hybridness of the modes is increased from 0.083 and 0.122 to reach a maximum value of 0.9985 and 0.9774 for the quasi-TE and quasi-TM modes, respectively. However, if φ is further increased, the modal hybridness will be reduced from its maximum value. At $\varphi = 90°$, the

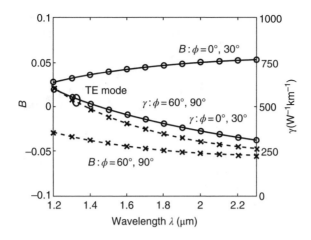

Figure 5.9   Variation of the birefringence B and nonlinearity γ for the quasi-TE mode with the wavelength at different φ. *Source*: Ref. [4]

modal hybridness of the quasi TE and quasi TM modes are 0.0913 and 0.114, respectively. The dependence of the hybridness on the rotation angle is very useful in designing highly tunable polarization conversion devices.

5.5.8   Effect of the Temperature

As shown in Figure 5.2, the ordinary n_o and extraordinary n_e refractive indices of the E7 material are influenced by the temperature variation that affects the modal properties of the proposed design. As the temperature T increases from 15 to 50°C, n_e of the E7 material decreases from 1.7096 to 1.6438 at the operating wavelength $\lambda = 1.55$ μm. However, n_o decreases slightly from 1.5034 to 1.5017 when the temperature changes from 15 to 35°C at $\lambda = 1.55$ μm, and then n_o increases from 1.5017 to 1.5089 when T increases from 35 to 50°C. Previously, Alkeskjold *et al.* [49] demonstrated a highly tunable PBG fiber, infiltrated with an optimized LC mixture having a large temperature gradient of the refractive indices at room temperature. In addition, Yuan *et al.* [50] studied the thermal tunability of bandgaps in microstructured polymer optical fibers infiltrated with LC. Therefore, the effect of the temperature variation on the modal properties of the suggested index guiding SGLC-PCF is also investigated.

The variations of the effective indices of the two polarized modes with the wavelength at different temperatures, from 15 to 35°C with a step of 5°C are evaluated, while the other parameters are fixed to $\Lambda = 2.0$ μm, $r_o = 0.5$ μm, $r = 0.9$ μm, and $\varphi = 90°$. At $\varphi = 90°$, ε_r of the E7 material is the diagonal of $[n_o^2, n_e^2, n_o^2]$. Therefore, ε_{yy} decreases with increasing the temperature, while ε_{xx} is nearly invariant. Therefore, the effective refractive index of the core region, and hence the effective index of the quasi-TM mode, decreases, as shown in Figure 5.10. However, the temperature variation at $\varphi = 90°$ has no effect on the effective index of the quasi-TE mode. This means that the birefringence of the suggested structure decreases with increasing the temperature from 15 to 35°C, as shown in Figure 5.11. Figure 5.11 shows the variation of the birefringence and effective mode area of the quasi TE mode with the wavelength at different temperatures from 15 to 35°C with a step of 5°C. At $T = 15°C$, the proposed design offers high birefringence of 0.045 at $\lambda = 1.55$ μm with circular holes only.

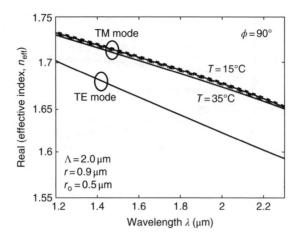

Figure 5.10 Variation of n_{eff} of the quasi-TE and quasi-TM modes with the wavelength at different temperatures T: 15, 20, 25, 30, and 35°C. *Source*: Ref. [3]

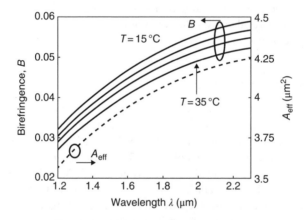

Figure 5.11 Variation of A_{eff} of the quasi-TE mode and birefringence with the wavelength at different temperatures T: 15, 20, 25, 30, and 35°C. *Source*: Ref. [3]

On the other hand, at $\varphi = 0°$, ε_r is the diagonal of $[n_e^2, n_o^2, n_o^2]$; therefore, ε_{xx} is affected by the temperature variation, while ε_{yy} is nearly constant. Consequently, the variation of the birefringence with the temperature at $\varphi = 0°$ is only due to the change in the n_{eff} of the quasi-TE mode. Through the range of φ from $0°$ to less than $45°$, ε_{xx} depends on n_e, while ε_{yy} relies on n_o. As a result, the variation of the birefringence with temperature is only due to the variation of n_{eff} of the quasi-TE mode. However, in the range of φ from $45°$ to $90°$, ε_{xx} depends on n_o, while ε_{yy} relies on n_e; therefore, ε_{yy} is affected by the temperature variation, while ε_{xx} is nearly constant. For this reason, the variation of birefringence with the temperature through the range of φ from $45°$ to $90°$ is only due to the change in the n_{eff} of the quasi-TM mode.

It is also evident from Figure 5.11 that the temperature has a slight effect on the effective mode area of the quasi-TE due to the slight effect of the temperature on ε_{xx}. However, the numerical results reveal that the effective mode area of the quasi-TM mode increases with increasing temperature. As the temperature increases from 15 to 35°C, ε_{yy} decreases from 2.9227 to 2.7021. Therefore, the index contrast seen by the quasi-TM mode decreases, and hence the mode confinement through the core region decreases.

5.5.9 *Elliptical SGLC-PCF*

The effect of the deformation of the infiltrated central circular hole into an elliptical hole is further studied. Here, a_0 and b_0 are the radii in the x- and y-directions of the elliptical hole as shown in the inset of Figure 5.12. The variation of the birefringence with the wavelength at different b_0 values, 0.35, 0.4, 0.45, and 0.5 μm is plotted in the figure, while the other parameters are kept constant at $\Lambda = 2.0$ μm, $r = 0.9$ μm, $a_0 = 0.5$ μm, $\varphi = 90°$, and $T = 25°C$. The variation of the birefringence of the soft glass PCF with a central air hole is also shown in the figure to understand the effect of the infiltration of the NLC in the central hole. It is clearly evident from the figure that the birefringence of the SGLC-PCF increases with increasing b_0. As b_0 increases from 0.35 to 0.5 μm, the birefringence increases from 0.03 to 0.042 at the operating wavelength $\lambda = 1.55$ μm. However, the birefringence of the soft glass PCF decreases with increasing the b_0 value. In addition, the birefringence of the SGLC-PCF is greater than that of the soft glass PCF with air holes which demonstrates the effect of the infiltration of NLC. The SGLC-PCF with circular holes ($a_0 = b_0 = 0.5$ μm) offers high birefringence of 0.042 at the operating wavelength $\lambda = 1.55$ μm. However, only birefringences of 9.2×10^{-3} and 3.1×10^{-4} are reported using soft glass PCF with a central elliptical hole ($a_0 = 0.5$ μm and $b_0 = 0.35$ μm) and a circular hole ($a_0 = b_0 = 0.5$ μm), respectively. The variation of the birefringence with the wavelength at different a_0 values, 0.35, 0.40, 0.45, and 0.5 μm is also investigated. In this study, the other parameters are kept constant at $\Lambda = 2.0$ μm, $r = 0.9$ μm, $b_0 = 0.5$ μm, $\varphi = 90°$ and $T = 25°C$. It is found that the effect of a_0 variation at constant b_0 is the same as that of b_0 variation at constant a_0.

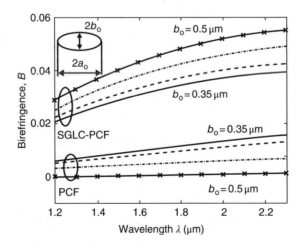

Figure 5.12   Variation of the birefringence of the SGLC-PCF and soft glass PCF with air holes with the wavelength at different b_0 values. *Source*: Ref. [3]

5.6   Experimental Results of LC-PCF

The SGLC-PCF [3] depends on the widely fabricated triangular lattice silica PCF [51]. Due to the dramatic improvement in the PCF fabrication process, more complex PCF structures can now be fabricated. PCF with a hole pitch of 2.3 μm or less have been fabricated successfully with 0.3 μm hole diameter [51]. Further, more complex PCF can be realized, such as V-shaped high birefringence PCFs [52]. The filling of PCFs with liquid or liquid-crystalline materials has already been demonstrated in the literature [8, 9, 44, 53, 54]. It should also be noted that PCFs with a hole pitch of 2.0 μm and hole diameter 1.0 μm or less (0.7–1.0 μm) infiltrated with NLC has been fabricated successfully by Woliński et al. [45]. In addition, the infiltration of only central holes has been suggested in Ref. [46] using similar filling technique with dimensions in the same region as our proposed PCF. Moreover, the two-photon direct-laser writing technique now offers full flexibility in selective filling of PCFs [47]. Huang et al. [54] reported a selective filling technique that depends on filling the large central hole faster than the smaller cladding holes. Hence, the PCF can be cleaved with the required length when only the central hole is filled with the infiltrated material.

5.6.1   Filling Temperature

The temperature of the LC material in the filling process plays an important role in the filling time, final alignment of the LC molecules, and propagation losses through the PCF. It has been found that the NLC of type E7 should be filled at its isotropic temperature to stabilize the LC alignment [48]. It then has to be cooled gradually, heated again to the isotropic temperature, and finally cooled. As a result, PCF with low insertion losses will be obtained [15].

5.6.2   Filling Time

The filling time depends on the contact angle θ, shown in Figure 5.13, between the LC and the inside wall of the PCF capillary. If $\theta < 90°$, the capillary forces will pull the liquid into the capillary; while if $\theta > 90°$, the capillary force will push the liquid out. The filling time of water in PCF has been calculated theoretically and verified experimentally by Nielsen et al. [53]. In this investigation [53], the water column inside the PCF capillary tube was exposed to gravitational force, capillary force, friction force due to the liquid viscosity, and force due to an overhead pressure. By taking into account these forces, the filling time and length relation can be obtained as reported in Ref. [53]. It should be noted that the LC has $\theta < 90°$, and so the LCs are pulled into the PCF capillary. The filling length of LMA-13 PCF with a hole diameter 4.3 μm versus filling time has been measured experimentally by Scolari [48]. In this study, two LCs, E7 (filled at the isotropic temperature) and MDA-00-3969 (filled at room temperature) are infiltrated by capillary forces. It was found that the filling speed of the E7 material is higher than for MDA-00-3969 due to the filling of the E7 type at its isotropic phase. In addition, the filling length shows a decaying exponential behavior as a function of time in both cases. Moreover, filling lengths of 11.2 and 5.1 cm are achieved for E7 and 5.1 cm for MDA-00-3969 after 3 h. The reported filling length is sufficient to be used in communication system devices such as polarization rotators and splitters. Further, the formation of long PCF can be followed by a cleaving process to create the required shorter length, as reported by Huang et al. [54].

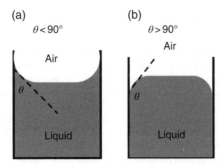

Figure 5.13   Contact angle θ between liquid and capillary wall for (a) $\theta < 90°$ and (b) $\theta > 90°$

Finally, coupling the light in this new type of PCF can be realized by splicing to a standard SMF and then launching the light from a laser source direct to the SMF [55]. In order to increase the coupling between the reported PCF and the SMF and reduce the coupling loss, tapered PCFs [56] or microtips [55] can be used. Coupling to small-core PCFs can also be realized with relatively high coupling efficiency using a bulk optics setup coupled to standard fiber patch cables, as reported in Ref. [57]. The introduced highly birefringent PCF can be used for polarization control in fiber optic sensors, precision optical instruments, and optical communication systems. Furthermore, the SGLC-PCF can be designed as a polarization rotator due to its high hybridness. Moreover, if two infiltrated dual NLC cores are introduced in the SGLC-PCF, a highly tunable and highly birefringent coupler can be produced. The suggested coupler will have a strong potential to offer a practical design for a short polarization splitter as well as a multiplexer–demultiplexer.

5.7 Summary

The chapter starts with brief overview of the LC-PCF followed by some background on LCs and soft glasses. Then, a novel design of high birefringence index-guiding SGLC-PCF is introduced and analyzed by the FVFDM [1]. The numerical results reveal that the SGLC-PCF has high birefringence, nonlinearity, and high tunability with temperature or an external electric field. It is also evident from the numerical results that the SGLC-PCF supports highly hybrid modes, which is very useful in designing polarization conversion devices, as will be described in Chapter 8. The high birefringence NLC-PCF is also very useful in designing PCF-based directional couplers and polarization splitters, as will be introduced in Chapter 9.

References

[1] Fallahkhair, A.B., Li, K.S., and Murphy, T.E. (2008) Vector finite difference modesolver for anisotropic dielectric waveguides. *J. Lightwave Technol.*, **26** (11), 1423–1431.
[2] Hameed, M., Obayya, S.S.A., Al-Begain, K. *et al.* (2009) Modal properties of an index guiding nematic liquid crystal based photonic crystal fiber. *Lightwave Technol. J.*, **27**, 4754–4762.
[3] Hameed, M.F.O. and Obayya, S.S.A. (2012) Modal analysis of a novel soft glass photonic crystal fiber with liquid crystal core. *Lightwave Technol. J.*, **30**, 96–102.
[4] Hameed, M.F.O., Obayya, S.S.A., and El-Mikati, H.A. (2011) Highly nonlinear birefringent soft glass photonic crystal fiber with liquid crystal core. *Photon. Technol. Lett. IEEE*, **23**, 1478–1480.

[5] Hu, C. and Whinnery, J.R. (1974) Losses of a nematic liquid-crystal optical waveguide. *J. Opt. Soc. Am.*, **64**, 1424–1432.

[6] Zografopoulos, D.C., Kriezis, E.E., and Tsiboukis, T.D. (2006) Photonic crystal-liquid crystal fibers for single-polarization or high-birefringence. *Opt. Express*, **14** (2), 914–925.

[7] Woliński, T.R., Szaniawska, K., Ertman, S. *et al.* (2006) Influence of temperature and electrical fields on propagation properties of photonic liquid crystal fibers. *Meas. Sci. Technol.*, **17** (50), 985–991.

[8] Fang, D., Yan, Q.L., and Shin, T.W. (2004) Electrically tunable liquid-crystal photonic crystal fiber. *Appl. Phys. Lett.*, **85** (12), 2181–2183.

[9] Haakestad, M.W., Alkeskjold, T.T., Nielsen, M. *et al.* (2005) Electrically tunable photonic bandgap guidance in a liquid-crystal-filled photonic crystal fiber. *IEEE Photon. Technol. Lett.*, **17** (4), 819–821.

[10] Haakestad, T. T. L. M. W., Nielsen, M. D., Engan, H. E., and Bjarklev, A. (2004) Electrically tunable fiber device based on a nematic liquid crystal filled photonic crystal fiber. Presented at the ECOC 2004, Stockholm, Sweden.

[11] Woliński, T.R., Tefelska, M., Milenko, K. *et al.* (2014) Propagation effects in a polymer-based photonic liquid crystal fiber. *Appl. Phys. A*, **115**, 569–574.

[12] Wei, L., Eskildsen, L., Weirich, J. *et al.* (2009) Continuously tunable all-in-fiber devices based on thermal and electrical control of negative dielectric anisotropy liquid crystal photonic bandgap fibers. *Appl. Opt.*, **48**, 497–503.

[13] Hameed, M.F.O., Obayya, S.S.A., and Wiltshire, R.J. (2010) Beam propagation analysis of polarization rotation in soft glass nematic liquid crystal photonic crystal fibers. *Photon. Technol. Lett. IEEE*, **22**, 188–190.

[14] Hameed, M.F.O. and Obayya, S.S.A. (2010) Analysis of polarization rotator based on nematic liquid crystal photonic crystal fiber. *Lightwave Technol. J.*, **28**, 806–815.

[15] Alkeskjold, T., Lægsgaard, J., Bjarklev, A. *et al.* (2004) All-optical modulation in dye-doped nematic liquid crystal photonic bandgap fibers. *Opt. Express*, **12**, 5857–5871.

[16] Alkeskjold, T.T. and Bjarklev, A. (2007) Electrically controlled broadband liquid crystal photonic bandgap fiber polarimeter. *Opt. Lett.*, **32** (12), 1707–1709.

[17] Wei, L., Alkeskjold, T.T., and Bjarklev, A. (2009) Compact design of an electrically tunable and rotatable polarizer based on a liquid crystal photonic bandgap fiber. *Photon. Technol. Lett. IEEE*, **21**, 1633–1635.

[18] Kumar, V.V.R.K., George, A.K., Reeves, W.H. *et al.* (2002) Extruded soft glass photonic crystal fiber for ultra-broadband supercontinuum generation. *Opt. Express*, **10**, 1520–1525.

[19] Fedotov, B., Sidorov-Biryukov, D.A., Ivanov, A.A. *et al.* (2006) Soft-glass photonic-crystal fibers for frequency shifting and white-light spectral superbroadening of femtosecond Cr:forsterite laser pulses. *J. Opt. Soc. Am. B*, **23**, 1471–1477.

[20] Friberg, S.R. and Smith, P.W. (1987) Nonlinear optical-glasses for ultrafast optical switches. *IEEE J. Quantum Electron.*, **23**, 2089–2094.

[21] Kumar, V., George, A.K., Knight, J.C., and Russell, P.St.J. (2003) Tellurite photonic crystal fibre. *Opt. Express*, **11**, 2641–2645.

[22] Leong, J. Y. Y. (2007) Fabrication and applications of lead-silicate glass holey fiber for 1-1.5 microns: nonlinearity and dispersion trade offs. PhD thesis. University of Southampton.

[23] Petropoulos, P., Monro, T.M., Ebendorff-Heidepriem, H. *et al.* (2003) Highly nonlinear and anomalously dispersive lead silicate glass holey fibres. *Opt. Express*, **11**, 3568–3573.

[24] Monro, T.M., West, Y.D., Hewak, D.W. *et al.* (2000) Chalcogenide holey fibres. *Electron. Lett.*, **36**, 1998–2000.

[25] van Eijkelenborg, M.A., Large, M.C.J., Argyros, A. *et al.* (2001) Microstructured polymer optical fibre. *Opt. Express*, **9**, 319–327.

[26] Mori, A., Shikano, K., Enbutsu, K. *et al.* (2004) *1.5 μm band zero-dispersion shifted tellurite photonic crystal fibre with a nonlinear coefficient of 675 W^{-1} km^{-1}.* ECOC 2004, 30th European Conference on Optical Communication, September 5–9, 2004, Stockholm, Sweden, p. Th3.3.6.

[27] Leong, J. Y. Y., Petropoulos, P., Asimakis, S. *et al.* (2005) A lead silicate holey with $\gamma = 1860$ $W^{-1}km^{-1}$ at 1550nm. Presented at the OFCs 2005, Anaheim.

[28] Leong, J. Y. Y. (2007) Fabrication and applications of lead-silicate glass holey fiber for 1–1.5 microns: nonlinearity and dispersion trade offs. PhD thesis. Faculty of Engineering, Science and Mathematics Optoelectronics Research Centre, University of Southampton.

[29] Ebendorff-Heidepriem, P.P.H., Asimakis, S., Finazzi, V.R. *et al.* (2004) Bismuth glass holey fibres with high nonlinearity. *Opt. Express*, **12**, 5082–5087.

[30] Li, J., Wu, ST., Brugioni, S. *et al.* (2005) Infrared refractive indices of liquid crystals. *J. Appl. Phys.*, **97** (7), 073501–073501-5.

[31] Ren, G., Shum, P., Yu, X. *et al.* (2008) Polarization dependent guiding in liquid crystal filled photonic crystal fibers. *Opt. Commun.*, **281**, 1598–1606.

[32] Acharya, B.R., Baldwin, K.W., Rogers, J.A. *et al.* (2002) In-fiber nematic liquid crystal optical modulator based on in-plane switching with microsecond response time. *Appl. Phys. Lett.*, **81** (27), 5243–5245.

[33] Haxha, S. and Ademgil, H. (2008) Novel design of photonic crystal fibers with low confinement losses, nearly zero ultra-flatted chromatic dispersion, negative chromatic dispersion and improved effective mode area. *J. Opt. Commun.*, **281**, 278–286.

[34] Obayya, S.S.A., Haxha, S., Rahman, B.M.A. *et al.* (Aug. 2003) Optimization of optical properties of a deeply-etched semiconductor optical modulator. *J. Lightwave Technol.*, **21** (8), 1813–1819.

[35] Agrawal, G.P. (1997) *Fiber-Optic Communication Systems*, 2nd edn. John Wiley & Sons, Inc., Rochester.

[36] Mortensen, N.A. and Folkenberg, J.R. (2003) Low-loss criterion and effective area considerations for photonic crystal fibres. *J. Opt. A: Pure Appl. Opt.*, **5**, 163–167.

[37] Agrawal, G.P. (1995) *Nonlinear Fibre Optics*, 2nd edn, Academic Press, Inc., San Diego.

[38] Okuno, T., Onishi, M., and Kashiwada, T. *et al.* (1999) Silicabased functional fibres with enhanced nonlinearity and their applications. *IEEE J. Sel. Top. Quantum*, **5**, 1385–1391.

[39] Lee, J.H., Belardi, W., and Furusawa, K. *et al.* (2003) Four-wave mixing based 10-Gb/s tunable wavelength conversion using a holey fibre with a high SBS threshold. *IEEE Photon. Technol. Lett.*, **15**, 440–442.

[40] Dakin, J.P. and Brown, R.G.W. (2006) *Handbook of Optoelectronics*, vol. **II**, Taylor & Francis, New York.

[41] Ebendorff-Heidepriem, H., Petropoulos, P., Finazzi, V. *et al.* (2004) *Highly nonlinear bismuthoxide-based glass holey fibre*, Optical Fiber Communication Conference, 2004 (OFC 2004), Los Angeles, California, p. ThA4.

[42] Kiang, K.M., Frampton, K., Monro, T.M. *et al.* (2002) Extruded singlemode non-silica glass holey optical fibres. *Electron. Lett.*, **38**, 546–547.

[43] Yue, Y., Kai, G., Wang, Z. *et al.* (2007) Highly birefringent elliptical-hole photonic crystal fiber with squeezed hexagonal lattice. *Opt. Lett.*, **32**, 469–471.

[44] Xiao, L., Jin, W., Demokan, M.S. *et al.* (2005) Fabrication of selective injection microstructured optical fibers with a conventional fusion splicer. *Opt. Express*, **13** (22), 9014–9022.

[45] Woliński, T.R., Szaniawska, K., Bondarczuk, K. *et al.* (2005) Propagation properties of photonic crystal fibers filled with nematic liquid crystals. *Opto-Electron. Rev.*, **13**, 177–182.

[46] Woliński, T.R., Czapla, A., Ertman, S. *et al.* (2007) Tunable highly birefringent solid-core photonic liquid crystal fibers. *Opt. Quantum Electron.*, **39**, 1021–1032.

[47] Vieweg, M., Gissibl, T., Pricking, S. *et al.* (2010) Ultrafast nonlinear optofluidics in selectively liquid-filled photonic crystal fibers. *Opt. Express*, **18**, 25232–25240.

[48] Scolari, L. (2009) Liquid crystals in photonic crystal fibers: fabrication, characterization and devices, PhD, Department of Photonics Engineering, Technical University of Denmark.

[49] Alkeskjold, J.L.T.T., Bjarklev, A., Hermann, D.S. *et al.* (2006) Highly tunable large-core single-mode liquid-crystal photonic bandgap fiber. *Appl. Opt.*, **45**, 2261–2264.

[50] Yuan, L.W.W., Alkeskjold, T.T., Bjarklev, A. and Bang, O. (2009) Thermal tunability of photonic bandgaps in liquid crystal infiltrated microstructured polymer optical fibers. *Opt. Express*, **17**, 19356–19364.

[51] Knight, J.C., Birks, T.A., Russell, P.St.J., and Atkin, D.M. (1996) All-silica single-mode optical fiber with photonic crystal cladding. *Opt. Lett.*, **21**, 1547–1549.

[52] Wojcik, J., Mergo, P., Makara, M. *et al.* (2010) V type high birefringent PCF fiber for hydrostatic pressure sensing. *Photon. Lett. Poland*, **2**, 10–12.

[53] Nielsen, K., Noordegraaf, D., Sørensen, T. *et al.* (2005) Selective filling of photonic crystal fibers'. *J. Opt. A: Pure Appl. Opt.*, **7** (8), L13–L20.

[54] Huang, Y., Xu, Y., and Yariv, A. (2004) Fabrication of functional microstructured optical fibers through a selective-filling technique. *Appl. Phys. Lett.*, **85** (22), 5182–5184.

[55] Xiao, W.J.L.M., Demokan, M.S., Ho, H.L., Tam, H.Y., Ju, J. and Yu, J. (2006) Photopolymer microtips for efficient light coupling between single-mode fibers and photonic crystal fibers. *Opt. Lett.*, **30**, 1791–1793.

[56] Wadsworth, A.W.W., Leon-Saval, S. and Birks, T. (2005) Hole inflation and tapering of stock photonic crystal fibres. *Opt. Express*, **13**, 6541–6549.

[57] NKT Photonics Application Note (2009) *Fiber Handling Stripping, Cleaving and Coupling*, vol. **1.0**, Birkerød, Denmark.

6

Full-Vectorial Beam Propagation Method

6.1 Introduction

Numerical simulations play an important role in the design and modeling of guided wave optoelectronic devices. There are various modeling methods in which not only a full-vector model but also an approximate scalar model is used. In this chapter, an overview of the beam propagation methods (BPMs) [1–3] is given followed by the formulation of the full-vectorial finite-difference BPM (FVFD-BPM) [4], which is employed in this book to study the propagation through the axial direction of the nematic liquid crystal photonic crystal fiber (NLC-PCF)-based devices. In addition, the numerical precision of the FVFD-BPM is demonstrated through analysis of a soft glass PCF coupler infiltrated by NLC material [5]. It is found that the FVFD-BPM numerical results are in excellent agreement with those calculated by the finite-difference method (FDM) [6] and mode-matching (MM) method [7].

6.2 Overview of the BPMs

The BPM has been one of the most popular techniques for modeling and simulating optical guided-wave devices. The major concept of the BPM is the development of a formula that permits the propagation of an initial field distribution along the axial direction in steps of sufficiently small length as, shown in Figure 6.1 [8]. Early publications focused on the solution of the scalar paraxial wave equation by means of the fast Fourier transform (FFT) [9]. However, the formulation of the FFT-BPM is derived under the assumption that the refractive index difference in the transverse direction is very small; therefore, the FFT-BPM cannot be applied to structures with large index discontinuities. In addition, the FFT-BPM can only be used to study the scalar wave propagation and the vectorial properties of the guided wave cannot be described.

Computational Liquid Crystal Photonics: Fundamentals, Modelling and Applications, First Edition.
Salah Obayya, Mohamed Farhat O. Hameed and Nihal F.F. Areed.
© 2016 John Wiley & Sons, Ltd. Published 2016 by John Wiley & Sons, Ltd.

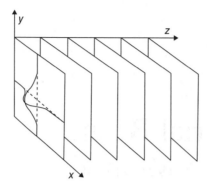

Figure 6.1 Propagation of an initial field distribution along the axial direction

The FDM was first introduced by Hendow and Shakir [10] to solve the paraxial scalar wave equation through a cylindrically symmetric structure. Then, Chung and Dagli [11] introduced the FD-BPM to the Cartesian coordinate system. It is worth noting that the introduction of scalar FDs within the framework of the BPM led to more refined schemes with increased efficiency and broader applicability [12, 13]. However the most serious drawback of the scalar FD scheme was the complete absence of the vector characteristics that are inherent in light propagation through inhomogeneous and/or anisotropic media. This was partly removed by the semi-vectorial BPM (SVBPM) that distinguishes two orthogonal but otherwise completely uncoupled states of polarization; transverse electric (TE) and transverse magnetic (TM) [2, 14]. The first approach to consider the vectorial nature of light was presented in Ref. [15] via an FD Crank–Nicolson scheme. However, the applicability of this approach [15] was restricted to planar straight waveguides. Other vector BPMs (VBPMs) have been developed based on the FDM [4, 16, 17].

Due to its numerical efficiency and versatility, some FV-BPM algorithms have been formulated based on the FEM [18–22]. Polstyanko *et al.* [18] investigated the full-vector finite-element BPM (FE-BPM) with electric field formulation. However, Obayya *et al.* [22] formulated the vector FE-BPM using the TM field components. In Ref. [23], an efficient vector FE-BPM for transverse anisotropic material was reported in terms of the TM field components with perfect matched layer boundary conditions and a wide-angle approximation.

In this chapter, the FVFD-BPM [4] is used to simulate the evolution of the magnetic field of an input optical field upon propagation through anisotropic media. In particular, it [4] is used to study the propagation through LC-based PCFs. In order to represent the orientation of the anisotropic LC, the full dielectric tensor must be considered in the BPM.

6.3 Formulation of the FV-BPM

Starting from Maxwell's equations, as described in Chapter 4, the vector wave equation for the magnetic field vector, H, can be derived as follows:

$$\nabla^2 H + k_o^2 \bar{\bar{\varepsilon}}_r H = -\bar{\bar{\varepsilon}}_r^{-1} \nabla \bar{\bar{\varepsilon}}_r \times (\nabla \times H). \tag{6.1}$$

Here, k is the free space wave number $k^2 = \omega^2 \mu_0 \varepsilon_0$, ω is the angular frequency, μ_0 is the permeability of free space, ε_0 is the permittivity of free space , and $\bar{\bar{\varepsilon}}_r$ is the permittivity tensor of the waveguide material which is given by [6]

$$\bar{\bar{\varepsilon}} = \varepsilon_0 \bar{\bar{\varepsilon}}_r = \varepsilon_0 \begin{pmatrix} \varepsilon_{xx} & \varepsilon_{xy} & 0 \\ \varepsilon_{yx} & \varepsilon_{yy} & 0 \\ 0 & 0 & \varepsilon_{zz} \end{pmatrix}. \tag{6.2}$$

The transverse component of the magnetic vector wave equation (6.1) can be expressed as follows:

$$\nabla^2 H_t + k_o^2 \bar{\bar{\varepsilon}}_{rt} H_t = -\bar{\bar{\varepsilon}}_{rt}^{-1} \nabla_t \bar{\bar{\varepsilon}}_{rt} \times \left(\nabla_t \times H_t \right). \tag{6.3}$$

Here, the subscript "t" stands for the transverse components and ε_{rt} is the transverse components of the dielectric tensor, and it can be defined such that

$$\bar{\bar{\varepsilon}}_{rt} = \begin{pmatrix} \varepsilon_{xx} & \varepsilon_{xy} \\ \varepsilon_{yx} & \varepsilon_{yy} \end{pmatrix} \tag{6.4}$$

6.3.1   Slowly Varying Envelope Approximation

The slowly varying approximation [4, 8] assumes that, since the simulation follows the propagation of light in the structure, the optical field can be defined in terms of its envelope and rapid phase components, as shown in Figure 6.2, that is,

$$H_t = \hat{H}_t e^{-jkn_0 z}. \tag{6.5}$$

Here, $\hat{H}_t$ is the envelope of the TM field component H_t, and n_0 is the reference index that is used to satisfy the slowly varying envelope approximation. In this chapter, n_0 is taken as the effective index of the fundamental mode launched at the input waveguide.

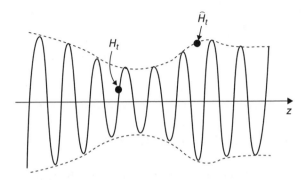

Figure 6.2   Envelope and rapid phase of an optical field

6.3.2   Paraxial and Wide-Angle Approximation

After substituting Eq. (6.5) in Eq. (6.3), the *Pade'* approximation of order zero [4] is used to decrease the time cost and required memory for the simulation. In this approximation, the second derivative $\partial^2 \hat{H}_t / \partial z^2$ is set to zero, which is accurate for most photonic guided-wave devices due to the slow variation of the envelope of the magnetic field in the z-direction. However, this approximation is not accurate in the case of structures that guide the light at a large angle to the assumed propagation direction. Therefore, a higher order of $\partial^2 / \partial z$ is required, which is called the higher Padé orders approximation (wide-angle situations). However, the memory requirement and computational time for the simulation are increased.

During this work, the paraxial approximation of order zero is used successfully for the beam propagation study for the NLC-PCF devices. In this case, after some algebraic manipulation, one can obtain the following two equations from the transverse wave equation (6.3):

$$j \frac{\partial \hat{H}_x}{\partial z} = B_{xx} \hat{H}_x + B_{xy} \hat{H}_y. \tag{6.6}$$

$$j \frac{\partial \hat{H}_y}{\partial z} = B_{yy} \hat{H}_y + B_{yx} \hat{H}_x. \tag{6.7}$$

The differential operators in (6.6) and (6.7), B_{xx}, B_{xy}, B_{yx}, and B_{yy} are also approximated using the FDM [6] and are defined such that

$$B_{xx} \hat{H}_x = \frac{1}{2kn_0} \left\{ \frac{\partial^2 \hat{H}_x}{\partial x^2} + \frac{\varepsilon_{yy}}{\varepsilon_{zz}} \frac{\partial^2 \hat{H}_x}{\partial y^2} + \frac{\varepsilon_{yx}}{\varepsilon_{zz}} \frac{\partial^2 \hat{H}_x}{\partial y \partial x} + k^2 \left(\varepsilon_{yy} - n_0^2 \right) \hat{H}_x \right\}. \tag{6.8}$$

$$B_{xy} \hat{H}_y = \frac{1}{2kn_0} \left\{ \left(1 - \frac{\varepsilon_{yy}}{\varepsilon_{zz}} \right) \frac{\partial^2 \hat{H}_y}{\partial x \partial y} - \frac{\varepsilon_{yx}}{\varepsilon_{zz}} \frac{\partial^2 \hat{H}_y}{\partial x^2} - k^2 \varepsilon_{yx} \hat{H}_y \right\}. \tag{6.9}$$

$$B_{yx} \hat{H}_x = \frac{1}{2kn_0} \left\{ \left(1 - \frac{\varepsilon_{xx}}{\varepsilon_{zz}} \right) \frac{\partial^2 \hat{H}_x}{\partial y \partial x} - \frac{\varepsilon_{xy}}{\varepsilon_{zz}} \frac{\partial^2 \hat{H}_x}{\partial y^2} - k^2 \varepsilon_{xy} \hat{H}_x \right\}. \tag{6.10}$$

$$B_{yy} \hat{H}_y = \frac{1}{2kn_0} \left\{ \frac{\partial^2 \hat{H}_y}{\partial y^2} + \frac{\varepsilon_{xx}}{\varepsilon_{zz}} \frac{\partial^2 \hat{H}_y}{\partial x^2} + \frac{\varepsilon_{xy}}{\varepsilon_{zz}} \frac{\partial^2 \hat{H}_y}{\partial x \partial y} + k^2 \left(\varepsilon_{xx} - n_0^2 \right) \hat{H}_y \right\}. \tag{6.11}$$

Using the Crank–Nicholson algorithm, Eqs. (6.6) and (6.7) can be rewritten in the following forms [4]:

$$\left[1 + j \Delta z \alpha B_{xx} \right] \hat{H}_x^{L+1} = \left[1 - j \Delta z (1 - \alpha) B_{xx} \right] \hat{H}_x^L - j \Delta z B_{xy} \hat{H}_y^L. \tag{6.12}$$

$$\left[1 + j \Delta z \alpha B_{yy} \right] \hat{H}_y^{L+1} = \left[1 - j \Delta z (1 - \alpha) B_{yy} \right] \hat{H}_y^L - j \Delta z B_{yx} \hat{H}_x^L. \tag{6.13}$$

Equations (6.12) and (6.13) for the TM fields can be solved by an iterative procedure to get the required magnetic fields. In this book, the required magnetic fields are calculated using an iterative Gaussian elimination method provided through a built-in MATLAB function (matrix left division\operator). The magnetic fields, $\widehat{H}_x^{L+1}$ and $\widehat{H}_y^{L+1}$ at a distance $L+1$ in the z-direction are obtained from the previous magnetic fields, $\widehat{H}_x^L$ and $\widehat{H}_y^L$, respectively at a distance L in the z-direction. The parameter α is introduced to control the scheme used to solve the FD equations. Through all simulations in this book, α is chosen within the range, $0.5 \leq \alpha \leq 1$, where the beam propagation scheme is unconditionally stable [24]. In addition, the propagation step Δz is taken as $1.0\,\mu m$.

6.4   Numerical Assessment

Due to their different uses in communication systems, the directional couplers [25–28] and their applications to the polarization splitter and the multiplexer–demultiplexer (MUX–DEMUX) have attracted the interest of many researchers in recent years. In this section, the numerical precision of the FVFD-BPM [4] is demonstrated through analysis of a novel design of directional coupler based on NLC-PCF [5].

6.4.1   Overview of Directional Couplers

The fiber coupler is one of the most important components in the optical communication systems. Mangan *et al.* [25] have shown that it is possible to use the PCF as an optical fiber coupler, which has some advantages over conventional optical fiber couplers. PCF couplers can be easily realized by simply introducing two adjacent defects in the PCF. In addition, they have short coupling length and better flexibility design. For these reasons, several studies have been reported for the dual-core PCF couplers [25–28]. The coupling characteristics of two different dual core PCF couplers were evaluated [26] showing that it is possible to realize significantly shorter MUX–DEMUX PCFs, compared to conventional optical fiber coupler. Additionally, a novel design of PCF splitter based on dual-core PCF with polarization-independent propagation characteristics was reported by Florous *et al.* [27]. They also [29] proposed a polarization-independent splitter, based on a highly birefringent dual-core PCF, which allows wavelength multiplexing at 1.3 and 1.55 µm. Furthermore, a polarization-independent splitter based on index guiding all silica-based PCF was suggested. See Ref. [30].

In this chapter, the coupling characteristics of a novel design of directional coupler [5] based on the reported soft glass NLC-PCF [31] are studied by the FVFD-BPM [4]. As mentioned before, the NLC-PCF is made from soft glass and NLC of types SF57 (lead silica) and E7, respectively. The refractive index of the SF57 material is greater than the ordinary and extraordinary refractive indices of the E7 material. Therefore, propagation through the NLC-PCF coupler occurs by modified total internal reflection. Further, the infiltration of the NLC [32–34] increases the birefringence between the two fundamental polarized modes in the proposed coupler. The soft glass background [35, 36] provides optical properties that cannot be obtained with silica, such as high refractive index, high rare-earth solubility, and mid-infrared transmission. In this study, the effects of the coupler geometrical parameters, rotation angle of the director of the NLC and temperature on the coupling characteristics of the reported NLC-PCF coupler are investigated in detail. The crosstalk (CT) and beam propagation study for the NLC-PCF coupler are also included.

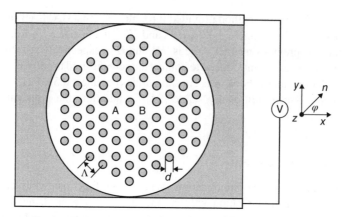

Figure 6.3 Cross section of the NLC-PCF coupler sandwiched between two electrodes and surrounded by silicon oil. The director of the NLC is shown at the right. *Source*: Ref. [5]

6.4.2   Design of the NLC-PCF Coupler

Figure 6.3 shows a cross section of the proposed NLC-PCF coupler whose cladding holes have been infiltrated with an NLC of type E7. All the holes have the same diameter d and are arranged with a hole pitch $\Lambda = 2.0\,\mu$m. The separation between the centers of the two cores, A and B, shown in Figure 6.3 is equal to $\sqrt{3}\Lambda$. In addition, the refractive index of the SF57 material and the ordinary and extraordinary refractive indices of the E7 material are fixed to 1.802, 1.5024, and 1.6970, respectively, at the operating wavelength $\lambda = 1.55\,\mu$m. The rotation angle, φ of the director of the NLC and the temperature are taken as 0 and 25°C, respectively.

The coupling length, defined as the minimum longitudinal distance at which maximum power is transferred from one core to another, is one of the important characteristics of directional couplers. The coupling length L_C at a given wavelength λ can be obtained using the difference between the effective indices $n_{\text{eff\_e}}$ and $n_{\text{eff\_o}}$ of the even and odd modes, respectively

$$L_C = \frac{\lambda}{2\left(n_{\text{eff\_e}} - n_{\text{eff\_o}}\right)}. \tag{6.14}$$

The effective indices of the even mode $n_{\text{eff\_e}}$ and odd mode $n_{\text{eff\_o}}$ of the NLC-PCF coupler are evaluated by the FVFDM [6] with perfect matched layer boundary conditions. Figure 6.4 shows the three-dimensional mode field profile of the fundamental component H_y of the even and odd TE modes at the operating wavelength $\lambda = 1.55\,\mu$m and with d/Λ of 0.7.

6.4.3   Effect of the Structural Geometrical Parameters

The effect of the coupler geometrical parameters on the coupling length of the NLC-PCF coupler [5] is investigated in detail. The d/Λ ratio is the first parameter to be studied. Figure 6.5 shows the variation of the coupling length for the two polarized modes of the NLC-PCF coupler

(a)   (b)

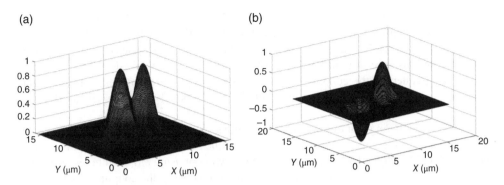

Figure 6.4 Three-dimensional mode field profile of the fundamental component H_y of the (a) even and (b) odd TE modes at the operating wavelength $\lambda = 1.55\,\mu m$

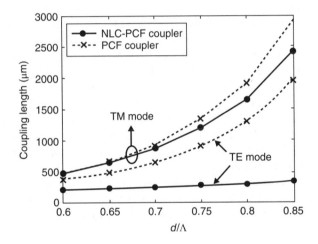

Figure 6.5 Variation of the coupling length of the soft glass NLC-PCF coupler and conventional silica PCF coupler with air holes for the two polarized modes with the d/Λ ratio. *Source*: Ref. [5]

with the d/Λ ratio. The hole pitch, rotation angle of the director of the NLC, temperature and wavelength are fixed at 2.0 μm, 0°, 25°C, and 1.55 μm, respectively. It is observed from Figure 6.5 that the coupling lengths of the two polarized modes increase with increasing d/Λ ratio at constant hole pitch Λ. As the d/Λ ratio increases at constant Λ, the soft glass bridge between the two cores of the NLC-PCF coupler decreases. Therefore, the distance traveled by the modes to transfer between the two cores, and hence the coupling length of the two polarized modes increase. As the d/Λ ratio increases from 0.6 to 0.85, the coupling length for the two polarized modes increases from 220.17 and 481.367 μm to 349.1 and 2421.9 μm for the quasi-TE and quasi-TM modes, respectively. In addition, it is revealed from Figure 6.5 that the coupling length for the quasi-TE mode is shorter than that for the quasi-TM mode.

The variation in the coupling length for the two polarized modes of the conventional silica PCF coupler with air holes is also shown in Figure 6.5. It should be noted that the index contrast between the core and cladding regions of the conventional PCF coupler [26] is greater than the index contrast of the NLC-PCF coupler. Therefore, the two fundamental polarized modes are more confined in the core regions of the conventional PCF coupler than in the core regions of the NLC-PCF coupler [5]. As a result, the modes of the conventional PCF coupler travel a longer distance to transfer between the two cores than the modes of the NLC-PCF coupler. Consequently, the coupling lengths for the two polarized modes of the soft glass NLC-PCF coupler are shorter than those of the conventional PCF coupler, as shown in Figure 6.5.

It is also worth noting that the birefringence, defined as the difference between the effective indices of the quasi-TE and -TM modes, is small for the conventional air holes PCF couplers, while the NLC-PCF coupler has high birefringence without using elliptical holes or two bigger holes in the first ring [37]. Therefore, the difference between the coupling lengths for the quasi-TE and -TM modes of the NLC-PCF coupler at a given d/Λ ratio is greater than that of the conventional PCF coupler, as can be seen from Figure 6.5.

The form birefringence [28] is defined as the ratio of $(L_{cTM} - L_{cTE})$ to L_{cTM} where L_{cTM} and L_{cTE} are the coupling lengths of the quasi-TM and -TE modes, respectively. Figure 6.6 shows the variation of the form birefringence of the soft glass NLC-PCF coupler and conventional silica PCF coupler with air holes with the d/Λ ratio while the hole pitch, rotation angle of the director of the NLC, temperature, and wavelength are fixed to 2 μm, 0°, 25°C, and 1.55 μm, respectively. It is revealed from the figure that the form birefringence of the conventional PCF coupler and NLC-PCF coupler increases with increasing d/Λ ratio. As the d/Λ ratio increases from 0.6 to 0.85, the form birefringence increases from 54.26 and 19.29% to 85.59 and 32.66% for the NLC-PCF coupler and the conventional PCF coupler, respectively. It can be seen from Figure 6.6 that the form birefringence of the NLC-PCF coupler is approximately 2.5 times that of the conventional PCF coupler. In addition, the form birefringence values indicate that the NLC-PCF coupler has strong polarization dependence; therefore, it can be used as a polarization splitter and its polarization dependence is stronger than those splitters presented in Refs. [28, 38, 39]. To do so, the NLC-PCF coupler can be carefully designed to obtain an integer ratio between the coupling lengths of the TE and TM modes at which the two different polarization

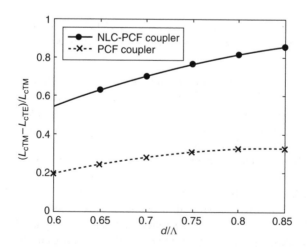

Figure 6.6 Variation of the form birefringence of the soft glass NLC-PCF coupler and conventional silica PCF coupler with air holes for the two polarized modes with the d/Λ ratio. *Source*: Ref. [5]

modes will exit at different cores. However, the conventional silica PCF coupler [26] has low birefringence, and high birefringence can be realized by adjusting the size of the air holes around the region of the two cores [28, 38, 39], which increases the difference between the coupling lengths for the two polarized modes. In Ref. [38], the two identical cores are formed by combination of large and small air holes, which makes the two cores birefringent. Zhang and Yang [39] reported a polarization splitter based on two nonidentical cores with also a combination of large and small air holes. However, in Ref. [28], two elliptical cores are used to improve the polarization dependence of the conventional silica PCF coupler.

6.4.4 Effect of Temperature

The ordinary and extraordinary refractive indices of the NLC are influenced by temperature that affects the coupling characteristics of the NLC-PCF coupler [5]. Therefore, the effect of the variation of the ordinary and extraordinary refractive indices of the NLC with temperature on the coupling length is the next parameter to be considered. Figure 6.7 shows the variation of the coupling length for the two polarized modes with the temperature at two different φ, $0°$ and $90°$, while the other parameters are fixed at $\Lambda = 2.0\,\mu m$, $d/\Lambda = 0.6$, and $\lambda = 1.55\,\mu m$. It can be seen from the figure that the coupling length of the quasi-TE mode at $\varphi = 0°$ increases with increasing temperature and that of the quasi-TM mode is nearly constant. As the temperature T increases from 15 to 50°C, the coupling length of the quasi-TE mode at $\varphi = 0°$ increases from 212.3 to 270.89 μm. On the contrary, at $\varphi = 90°$ the coupling length of the quasi-TM mode increases with increasing the temperature while the coupling length of the quasi-TE modes is nearly constant. As T varies from 15 to 50°C, the coupling length of the quasi-TM mode at $\varphi = 90°$ varies from 231 to 299 μm. The dependence of the coupling length on the temperature can be explained by analyzing the dominant field components, E_x and E_y of the quasi-TE and -TM modes and the direction of the director of the NLC. . At $\varphi = 0°$, ε_r of

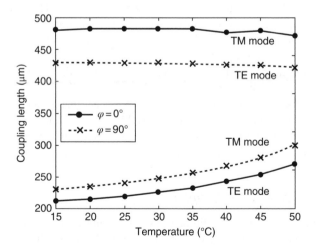

Figure 6.7 Variation of the coupling length for the two polarized modes of the NLC-PCF coupler with temperature at two different φ, $0°$, and $90°$ while the other parameters are fixed at $\Lambda = 2.0\,\mu m$, $d/\Lambda = 0.6$, and $\lambda = 1.55\,\mu m$. *Source*: Ref. [5]

the E7 material is the diagonal of $[\varepsilon_{xx}, \varepsilon_{yy}, \varepsilon_{zz}]$, where $\varepsilon_{xx} = n_e^2$, $\varepsilon_{yy} = n_o^2$, and $\varepsilon_{zz} = n_o^2$. As the temperature increases from 15 to 50°C, ε_{xx} decreases from 2.9227 to 2.7021, while ε_{yy} changes slightly from 2.2602 to 2.2768. Therefore, the index contrast for the quasi-TE modes increases with increasing the temperature, while the index contrast for the quasi-TM modes is nearly constant. Consequently, the confinement of the quasi-TE modes inside the core regions, and hence the distance needed by the quasi-TE modes to transfer between the two cores, increases with increasing temperature. Thus, the coupling length of the quasi-TE mode increases with increasing temperature, while the coupling length of the quasi-TM mode is nearly invariant, as shown in Figure 6.7.

On the other hand, at $\varphi = 90°$, ε_r of the E7 material is the diagonal of the $[\varepsilon_{xx}, \varepsilon_{yy}, \varepsilon_{zz}]$ where $\varepsilon_{xx} = n_o^2$, $\varepsilon_{yy} = n_e^2$, and $\varepsilon_{zz} = n_o^2$; therefore, ε_{yy} decreases with increasing temperature, while ε_{xx} is nearly constant. Consequently, the confinement of the quasi-TM modes inside the core regions, and hence the distance traveled by the quasi-TM modes to transfer between the two cores increases with increasing temperature. As a result, the L_C of the quasi-TM mode at $\varphi = 90°$ increases with increasing temperature, while the L_C of the quasi-TE mode is nearly constant, as observed in Figure 6.7.

6.4.5   Effect of the NLC Rotation Angle

As shown from Figure 6.7, the coupling length of the quasi-TE mode at $\varphi = 0°$ is shorter than that of the quasi-TM mode and the situation at $\varphi = 90°$ is reversed. At $\varphi = 0°$ and at $T = 15°C$, the coupling lengths for the quasi-TE and -TM modes are 212.3 and 480.6 μm, respectively. However, at $\varphi = 90°$ and $T = 15°C$, the values are 428.3 and 231 μm, respectively. The dependence of the coupling length on the rotation angle of the director of the NLC can also be explained by analyzing the dominant field components of the quasi-TE and -TM modes and the direction of the director of the NLC. At $\varphi = 0°$, the director of the NLC is parallel to E_x and perpendicular to E_y and the relative permittivity tensor ε_r of the E7 material is the diagonal of $[n_e^2, n_o^2, n_o^2]$. In this case, ε_{xx} is greater than ε_{yy}; therefore, the index contrast seen by the quasi-TE modes is less than that seen by the quasi-TM modes. Consequently, the quasi-TM modes are more confined in the core regions than the quasi-TE modes; therefore, the quasi-TM modes travel further than the quasi-TE modes to transfer from one core to another. As a result, the coupling length of the quasi-TE mode is shorter than that of the quasi-TM mode. On the other hand, at $\varphi = 90°$, the director of the NLC is parallel to E_y, while it is perpendicular to E_x and the relative permittivity tensor ε_r of the E7 material is diagonal of $[n_o^2, n_e^2, n_o^2]$. In this case, ε_{yy} is greater than ε_{xx} and the quasi-TE modes are more confined in the core regions than the quasi-TM modes; therefore, the quasi-TE modes travel further than the quasi-TM modes to transfer between the two cores, and hence the coupling lengths of the quasi-TE modes are longer than those of the quasi-TM modes.

6.4.6   Elliptical NLC-PCF Coupler

The effect of the deformation of the circular holes into elliptical holes on the performance of the suggested coupler is further studied. Here, a and b are the radii of the elliptical holes in the x- and y-directions, respectively, as shown in the inset of Figure 6.8. Figure 6.8 shows the variation of the coupling length of the two polarized modes of the NLC-PCF coupler [5] with the wavelength at different b values, 0.4, 0.5, and 0.6 μm, while the other parameters are taken as

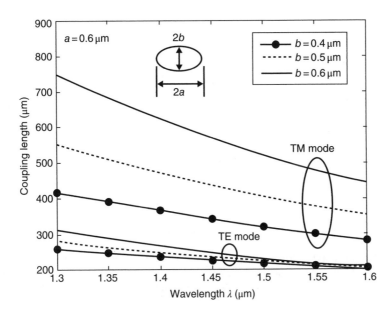

Figure 6.8 Variation of the coupling length for the two polarized modes of the NLC-PCF coupler with the wavelength at different b values, while the other parameters are invariant at $\Lambda=2.0\,\mu$m, $a=0.6\,\mu$m, $\varphi=0°$, and $T=25°$C where a and b are the radii of the elliptical holes in the x- and y-directions, respectively. *Source*: Ref. [5]

$\Lambda=2.0\,\mu$m, $a=0.6\,\mu$m, $\varphi=0°$, and $T=25°$C. It is observed from Figure 6.8 that the coupling lengths for the quasi-TE and -TM modes increase with increasing b. As b increases from 0.4 to 0.6 μm, the coupling length of the quasi-TE mode at $\lambda=1.55\,\mu$m increases slightly from 212.41 to 220.17 μm, while the coupling length of the quasi-TM mode increases from 300.14 to 481.37 μm. The variation of the form birefringence of the NLC-PCF coupler with the wavelength at different b values while the other parameters are invariant at $\Lambda=2.0\,\mu$m, $a=0.6\,\mu$m, $\varphi=0°$, and $T=25°$C is shown in Figure 6.9. It is seen that the form birefringence increases with increasing b with constant a. As b increases from 0.4 to 0.6 μm, the form birefringence at $\lambda=1.55\,\mu$m increases from 29.23 to 54.26%. In addition, the form birefringence and the coupling lengths for the two polarized modes decrease with increasing wavelength, as can be seen from Figures 6.8 and 6.9. At short wavelength, the two polarized modes are confined well within the core regions and as the wavelength increases the modes start spreading in the cladding region. Therefore, the modes at short wavelength travel further to transfer between the two cores than at long wavelength. As a result, the coupling length decreases with increasing wavelength, as shown in Figures 6.8 and 6.10.

Figure 6.10 shows the variation of the coupling length of the NLC-PCF coupler for the quasi-TE modes with the d/Λ ratio at different wavelengths, 1.3, 1.48, and 1.55, while the hole pitch, rotation angle of the director of the NLC and temperature are fixed to 2.0 μm, 90° and 25°C, respectively. As the d/Λ ratio increases from 0.6 to 0.85, the coupling lengths of the quasi TE modes increase from 654.28, 480.08, and 429.21 μm to 4229.92, 2354.49, and 1914.33 μm at $\lambda=1.3$, 1.48, and 1.55 μm, respectively. It may be observed that the coupling lengths at different wavelengths at a given d/Λ ratio can be tuned to design MUX–DEMUX

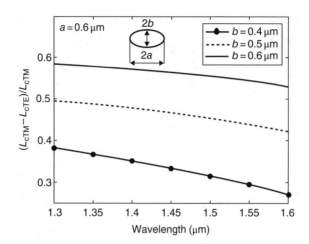

Figure 6.9 Variation of the form birefringence of the NLC-PCF coupler with the wavelength at different b values, while the other parameters are invariant at $\Lambda = 2.0\,\mu\text{m}$, $a = 0.6\,\mu\text{m}$, $\varphi = 0°$, and $T = 25°\text{C}$ where a and b are the radii of the elliptical holes in the x- and y-directions, respectively. *Source*: Ref. [5]

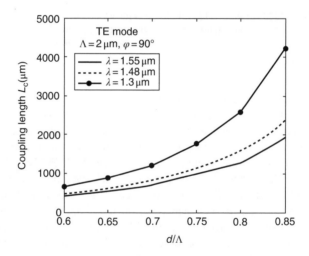

Figure 6.10 Variation of the coupling length for the quasi-TE modes of the NLC-PCF with the d/Λ ratio at different wavelengths 1.3, 1.48, and 1.55 μm. *Source*: Ref. [5]

couplers with short coupling lengths compared to conventional optical fiber couplers. To do this, the NLC-PCF coupler can be carefully designed to obtain an integer ratio between the coupling lengths of two different wavelengths for a given polarized mode at which the two different wavelengths will exit at different cores.

The variation of the coupling length with the wavelength at different a values, 0.4, 0.5, and 0.6 μm, is also investigated, while the other parameters are kept constant at $\Lambda = 2.0\,\mu\text{m}$, $b = 0.6\,\mu\text{m}$, $\varphi = 0°$, and $T = 25°\text{C}$. It is also found that the coupling lengths for the two polarized modes increase with increasing a at constant b.

6.4.7   Beam Propagation Analysis of the NLC-PCF Coupler

Finally, using the FVFD-BPM [4] and the mode-matching (MM) method [7], it is confirmed that the NLC-PCF coupler can couple the power between the two cores. In this study, the NLC-PCF coupler is considered with the following parameters: $\Lambda = 1.8\,\mu m$, $d/\Lambda = 0.53$, $\varphi = 90°$, $T = 25°C$, and $\lambda = 1.55\,\mu m$. As a directional coupler, this device supports even and odd modes that are continuously exchanging power during the propagation. In particular, maximum power transfer occurs at the coupling length L_C when the two modes become out of phase. Initially, at $z = 0$, the fundamental component H_y of the quasi-TE mode, of soft glass PCF with air holes obtained using the FVFDM [6], is launched into the left core of the NLC-PCF coupler [5]. This input field, in turn, starts to transfer to the right core of the coupler and at the coupling length L_C, the field is almost completely transferred to the right core. The calculated coupling lengths using the FVFD-BPM and MM method are 253 and 255 μm, respectively, which are in an excellent agreement with the 253.26 μm calculated by the FVFDM. The field distributions of the dominant field component H_y of the quasi-TE mode calculated by the FVFD-BPM are shown in Figure 6.11 at different waveguide sections z, 0, 100, 200, and 253 μm. It is evident from the figure that, at $z = 0$, the input field is launched into the left core and as the propagation distance increases the power in the right core increases and that in the left core decreases. At $z = 100\,\mu m$, the powers of the quasi-TE mode in the left and right cores of the coupler normalized to the input power are 0.6582 and 0.3404, respectively. However, at $z = 200\,\mu m$, the normalized powers are 0.1170 and 0.8807 in the left and right cores of the

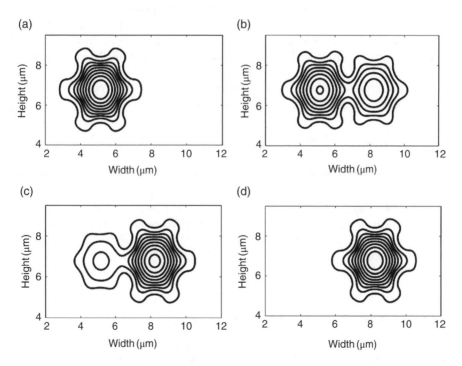

Figure 6.11   The field contour patterns for H_y of the quasi-TE mode at (a) $z = 0$, (b) $z = 100\,\mu m$, (c) $z = 200\,\mu m$, and (d) $z = 253\,\mu m$. *Source*: Ref. [5]

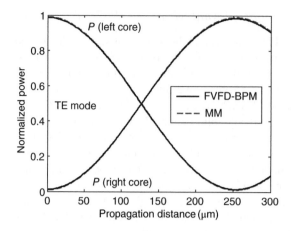

Figure 6.12 Evolution of the normalized powers along the propagation direction through the left and right cores of the proposed NLC-PCF coupler using the FVFD-BPM (*Source*: Ref. [4]) and the MM method (*Source*: Ref. [7]). *Source*: Ref. [5]

coupler, respectively. Finally, at $z = 253\,\mu$m, the power is almost completely transferred from the left core to the right core with a negligible fraction in the left core.

Figure 6.12 shows the variation of the normalized powers of the quasi-TE mode in the dual cores of the proposed NLC-PCF coupler along the propagation direction using the FVFD-BPM [4] and MM method [7]. As can be seen from the figure, almost complete power transfer occurs between the two cores of the NLC-PCF coupler. In addition, there is an excellent agreement between the normalized powers calculated by the FVFD-BPM and those obtained by the MM method.

The CT is a measure of the unwanted power, or the power of the left core, remaining at the end of the NLC-PCF coupler, which can be defined as follows:

$$CT = 10\log_{10} \frac{\text{Undesired normalized power in the left core A}}{\text{Desired normalized power in the right core B}} \text{ dB.} \qquad (6.15)$$

It is found that the CT calculated by the FVFD-BPM and the MM method are equal to −18.35 and −19.36 dB, respectively, at the corresponding coupling lengths, and it can be decreased further by increasing the hole pitch or the hole diameter. As the hole pitch increases from 1.8 to 2.2 μm, the CT decreases from −18.35 to −20.94 dB, while the d/Λ ratio is fixed at 0.53. In addition, the CT changes from −18.35 to −21.60 dB when the d/Λ ratio increases from 0.53 to 0.6, while the hole pitch is taken as 1.8 μm.

6.5 Experimental Results of LC-PCF Coupler

A directional coupler structure based on NLC selectively filled PCF was recently fabricated by Hu *et al.* [40]. In this study, only one hole next to the silica core has been filled with NLC of type 6CHBT (4-*trans-n*-hexyl-cyclohexyl-isothiocyanatobenzene). The fabricated device is compact of length 9 mm. Therefore, the presence of the bubbles in the filled channel is also greatly reduced due to shorter infiltration length. Through the NLC filling process, the PCF except for one air capillary is blocked by liquid glue. Then, the blocked side of the PCF is placed into the 6CHBT solution [40]. Therefore, the unblocked capillary will be infiltrated by the NLC via a capillary effect. The infiltration process can be confirmed by observing the other side of the PCF under a microscope. In order to control the temperature of the coupler, the LC-PCF is placed in an oven with temperature control accuracy of 0.1°C.

The refractive index of the NLC is higher than that of the silica background material. Therefore, the mode coupling between the NLC core and the silica core is more complicated than the coupling between two identical silica cores [40]. Further, higher-order modes through the NLC core will be coupled to the silica core guided modes. This occurs when their effective mode indices are equal. The coupling characteristics of the LC-PCF coupler have been studied experimentally [40]. It was found that the transmission spectra exhibited detectable notches for temperature sensing applications.

6.6　Summary

The chapter starts with a brief overview of BPMs. Then, the numerical assessment of the FVFD-BPM [4] is introduced through analysis of a highly tunable directional coupler [5] based on the index-guiding soft glass NLC-PCF. The analysis was made using the FVFDM [6], and confirmed by the FVFD-BPM [4]. The numerical results reveal that the NLC-PCF coupler has stronger polarization dependence than the conventional silica PCF couplers with air holes. Additionally, the NLC-PCF coupler has tremendous potential as a practical design for directional couplers, polarization splitters [41], and MUX–DEMUX [42] devices.

References

[1] Jose Patrocinio da Silva, H.E.H.-F. and Ferreira Frasson, A.M. (2003) Improved vectorial finite-element BPM analysis for transverse anisotropic media. *J. Lightwave Technol.*, **21** (2), 567–576.

[2] Huang, W.P., Xu, C., Chu, S.T., and Chaudhuri, S.K. (1992) The finite difference vector beam propagation method: analysis and assessment. *J. Lightwave Technol.*, **10**, 295–305.

[3] XU, C.L., Huang, W.P., Chrostowski, J., and Chaudhuri, S.K. (1994) A full-vectorial beam propagation method for anisotropic waveguides. *J. Lightwave Technol.*, **12** (11), 1926–1931.

[4] Huang, W.P. and Xu, C.L. (1993) Simulation of three-dimensional optical waveguides by a full-vector beam propagation method. *IEEE J. Quantum Electron.*, **29** (10), 2639–2649.

[5] Hameed, M.F.O., Obayya, S.S.A., Al Begain, K. *et al.* (2009) Coupling characteristics of a soft glass nematic liquid crystal photonic crystal fibre coupler. *IET Optoelectron.*, **3**, 264–273.

[6] Fallahkhair, A.B. *et al.* (2008) Vector finite difference modesolver for anisotropic dielectric waveguides. *J. Lightwave Technol.*, **26** (11), 1423–1431.

[7] Mustieles, F.J., Ballesteros, E., and Hernhdez-Gil, F. (1993) Multimodal analysis method for the design of passive TE/TM converters in integrated waveguides. *IEEE Photon. Technol. Lett.*, **5** (7), 809–811.

[8] Ma, F., Xu, C.L., and Huang, W.P. (April 1996) Wide-angle full vectorial beam propagation method. *IEE Proc. Optoelectron.*, **143** (2), 139–143.

[9] Feit, M.D. and Fleck, J.J.A. (1978) Light propagation in graded-index optical fibers. *Appl. Opt.*, **17** (24), 3990–3998.

[10] Hendow, S.T. and Shakir, S.A. (1986) Recursive numerical solution for nonlinear wave propagation in fibers and cylindrically symmetric systems. *Appl. Optics*, **25** (11), 1759–1764.

[11] Chung, Y. and Dagli, N. (1990) An assessment of finite difference beam propagation method. *IEEE J. Quantum Electron.*, **26** (8), 1335–1339.

[12] Yevick, D. and Hermansson, B. (1990) Efficient beam propagation techniques. *IEEE J. Quantum Electron.*, **26**, 109–112.

[13] Lee, P.C., Schulz, D., and Voges, E. (1992) Three-dimensional finite difference beam propagation algorithm for photonic devices. *J. Lightwave Technol.*, **10**, 1832–1838.

[14] Liu, P.L., Yang, S.L., and Yuan, D.M. (1993) The semivectorial beam propagation method. *IEEE J. Quantum Electron.*, **29**, 1205–1211.

[15] Clauberg, R. and von Allmen, P. (1991) Vectorial beam-propagation method for integrated optics. *Electron. Lett.*, **27** (8), 654–655.

[16] Chund, Y., Dagli, N., and Thylen, L. (1991) Explicit finite difference vectorial beam propagation method. *Electron. Lett.*, **27** (23), 2119–2121.

[17] Huang, W.P., Xu, C.L., and Chaudhuri, S.K. (1992) A finite-difference vector beam propagation method for three-dimensional waveguide structures. *IEEE Photon. Technol. Lett.*, **4**, 148–151.

[18] Polstyanko, S.V., Dyczij-Edlinger, R., and Lee, J.F. (1996) Full vectorial analysis of a nonlinear slab waveguide based on the nonlinear hybrid vector finite-element method. *Opt. Lett.*, **21**, 98–100.

[19] Montanari, E., Selleri, S., Vincetti, L., and Zoboli, M. (1998) Finite element full vectorial propagation analysis for three dimensional z-varying optical waveguides. *J. Lightwave Technol.*, **16**, 703–714.

[20] Koshiba, M., Tsuji, Y., and Hikari, M. (1999) Finite-element beam-propagation method with perfectly matched layers boundary conditions. *IEEE Trans. Magn.*, **35**, 1482–1485.

[21] Obayya, S.S.A., Rahman, B.M.A., and El-Mikati, H.A. (2000) New full vectorial numerically efficient propagation algorithm based on the finite element method. *J. Lightwave Technol.*, **18** (3), 409–415.

[22] Obayya, S.S.A., Rahman, B.M.A., and El-Mikati, H.A. (2000) Full vectorial finite element beam propagation method for nonlinear directional coupler devices. *IEEE J. Quantum Electron.*, **36**, 556–562.

[23] Patrocinio da Silva, J., Hugo, E., Figueroa, H., and Frasson, A.M.F. (2003) Improved vectorial finite-element BPM analysis for transverse anisotropic media. *J. Lightwave Technol.*, **21** (2), 567–576.

[24] Xu, C. and Huang, W.P. (1995) *Finite-Difference Beam Propagation Method for Guided-Wave Optics*. Progress In Electromagnetics Research, PIER 11, pp. 1–49.

[25] Mangan, B.J., Knight, J.C., Birks, T.A. *et al.* (2000) Experimental study of dual core photonic crystal fiber. *Electron. Lett.*, **36**, 1358–1359.

[26] Saitoh, K., Sato, Y., and Koshiba, M. (2003) Coupling characteristics of dual-core photonic crystal fiber couplers. *Opt. Express*, **11**, 3188–3195.

[27] Florous, N., Saitoh, K., and Koshiba, M. (2005) A novel approach for designing photonic crystal fiber splitters with polarization-independent propagation characteristics. *Opt. Express*, **13** (19), 7365–7373.

[28] Zhang, L. and Yang, C. (2004) Polarization- dependent coupling in twin-core photonic crystal fibers. *J. Lightwave Technol.*, **22** (5), 1367–1373.

[29] Florous, N., Saitoh, K., and Koshiba, M. (2006) Synthesis of polarization-independent splitters based on highly birefringent dual-core photonic crystal fiber platforms. *Photon. Technol. Lett.*, **18**, 1231–1233.

[30] Chen, M.Y. and Zhou, J. (2006) Polarization-independent splitter based on all-solid silica-based photonic-crystal fibers. *J. Lightwave Technol.*, **24**, 5082–5086.

[31] Hameed, M.F.O., Obayya, S.S.A., Al-Begain, K. *et al.* (2009) Modal properties of an index guiding nematic liquid crystal based photonic crystal fiber. *J. Lightwave Technol.*, **27**, 4754–4762.

[32] Wolinski, T.R., Szaniawska, K., Ertman, S. *et al.* (2006) Influence of temperature and electrical fields on propagation properties of photonic liquid crystal fibers. *Meas. Sci. Technol.*, **17** (50), 985–991.

[33] Zografopoulos, D.C., Kriezis, E.E., and Tsiboukis, T.D. (2006) Photonic crystal-liquid crystal fibers for single-polarization or high-birefringence. *Opt. Express*, **14** (2), 914–925.

[34] Ren, G., Shum, P., Yu, X. *et al.* (March 2008) Polarization dependent guiding in liquid crystal filled photonic crystal fibers. *Opt. Commun.*, **281**, 1598–1606.

[35] Kumar, V.V.R.K., George, A.K., Reeves, W.H. *et al.* (2002) Extruded soft glass photonic crystal fiber for ultra-broadband supercontinuum generation. *Opt. Express*, **10**, 1520–1525.

[36] Fedotov, B., Sidorov-Biryukov, D.A., Ivanov, A.A. *et al.* (2006) Soft-glass photonic-crystal fibers for frequency shifting and white-light spectral superbroadening of femtosecond Cr:forsterite laser pulses. *J. Opt. Soc. Am. B*, **23**, 1471–1477.

[37] Li, J., Duan, K., Wang, Y. *et al.* (2009) Design of a single-polarization single-mode photonic crystal fiber double-core coupler. *Opt. Int. J. Light Electron. Opt.*, **120** (10), 490–496.

[38] Zhang, L. and Yang, C. (2003) Polarization splitter based on photonic crystal fibers. *Opt. Express*, **11** (9), 1015–1020.

[39] Zhang, L. and Yang, C. (2004) A novel polarization splitter based on the photonic crystal fiber with nonidentical dual cores. *IEEE Photon. Technol. Lett.*, **16** (7), 1670–1672.

[40] Hu, D.J.J., Shum, P.P., Jun Long, L. *et al.* (2012) A compact and temperature-sensitive directional coupler based on photonic crystal fiber filled with liquid crystal 6CHBT. *IEEE Photon. J.*, **4**, 2010–2016.

[41] Hameed, M.F.O. and Obayya, S.S.A. (2009) Polarization splitter based on soft glass nematic liquid crystal photonic crystal fiber. *IEEE Photon. J.*, **1**, 265–276.

[42] Hameed, M.F.O., Obayya, S.S.A., and Wiltshire, R.J. (2009) Multiplexer–demultiplexer based on nematic liquid crystal photonic crystal fiber coupler. *Opt. Quantum Electron.*, **41**, 315–326.

7

Finite-Difference Time Domain Method

7.1 Introduction

Recently, optical elements with anisotropic dielectric properties, such as wave plates and liquid crystal materials, have achieved substantial practical use because of their capability to control the polarization of light waves. One of the most general, flexible, and powerful approaches for the analysis of arbitrary planar anisotropic structures is based on the finite-difference time domain (FDTD).

FDTD methods present an accurate and explicit approach to directly solving Maxwell's curl equations both in time and space. Further, one of the advantages of FDTD techniques is their ability to define arbitrary shapes, nonlinear behavior, and inhomogeneous properties. Using the FDTD technique offers a simple way to visualize real-time pictures of the electromagnetic wave [1, 2].

In 1966, Yee [3] was the first to present the idea of FDTD. The origin of Yee's algorithm was to simulating isotropic and linear materials. In the early 1980s, researchers began to explore the possibility of developing the FDTD algorithm to solve electromagnetic (EM) simulation problems having anisotropic media. In 1986, Choi and Hoefer [4] were the first to use the FDTD method for anisotropic media. After 7 years, Schneider and Hudson [5] modified the FDTD method to simulate materials with full permittivity and conductivity tensors. Also, the research of Hunsberger and Luebbers [6] was devoted to using the FDTD to analyze magnetized plasma. In 1999, Zhao proposed a three-dimensional (3D) algorithm to handle general anisotropic materials and, in 2002, Zhao extended Berenger's perfectly matched layer (PML) to solve general anisotropic materials [7]. In 2002, Moss and Teixeira studied the numerical dispersion in an anisotropic medium [8]. In 2004, uniaxial bianisotropic and magneto dielectric media were analyzed using FDTD by Akyurtlu and Werner [9] and Mosallaei and Sarabandi [10], respectively.

Computational Liquid Crystal Photonics: Fundamentals, Modelling and Applications, First Edition.
Salah Obayya, Mohamed Farhat O. Hameed and Nihal F.F. Areed.
© 2016 John Wiley & Sons, Ltd. Published 2016 by John Wiley & Sons, Ltd.

7.2   Numerical Derivatives

This section presents the basic idea of the FDTD method to solve Maxwell's differential equations. Using FD methods to solve differential equations is easily understood, more universal, and more applicable than other numerical or analytical approaches. The derivative of any arbitrary function $f(x)$ at any arbitrary point x_0 can be calculated by three forms: the forward difference, the backward difference, and the central difference. Equations (7.1), (7.2), and (7.3) show the calculated first-order derivation of $f(x)$ at x_0 using the three difference forms. In the same way, the second-order derivation of $f(x)$ at x_0 can be estimated as shown by Eq. (7.4).

The first derivative calculations using the forward difference form is as follows:

$$f'(x_0) \cong \frac{1}{\Delta x}\left(f\left(x_0 + \Delta x\right) - f\left(x_0\right)\right) \tag{7.1}$$

The first derivative calculations using the backward difference form is as follows:

$$f'(x_0) \cong \frac{1}{\Delta x}\left(f\left(x_0\right) - f\left(x_0 - \Delta x\right)\right) \tag{7.2}$$

The first derivative calculations using the central difference form is as follows:

$$f'(x_0) \cong \frac{1}{2\Delta x}\left(f\left(x_0 + \Delta x\right) - f\left(x_0 - \Delta x\right)\right) \tag{7.3}$$

The second derivative calculations using the central difference form is as follows:

$$
\begin{aligned}
f''(x_0) &\cong \frac{1}{\Delta x}\left(f'\left(x_0 + \frac{\Delta x}{2}\right) - f'\left(x_0 - \frac{\Delta x}{2}\right)\right) \\
&\cong \frac{1}{\Delta x}\left[\frac{f\left(x_0 + \Delta x\right) - f\left(x_0\right)}{\Delta x} - \frac{f\left(x_0\right) - f\left(x_0 - \Delta x\right)}{\Delta x}\right] \\
&\cong \left[\frac{f\left(x_0 + \Delta x\right) - 2f\left(x_0\right) + f\left(x_0 - \Delta x\right)}{\left(\Delta x\right)^2}\right]
\end{aligned}
\tag{7.4}
$$

The truncation error is defined as the approximation error between the accurate and approximate derivative calculations and is calculated by Taylor's expansion analyses. As indicated in Ref. [11], the errors for the forward difference, backward difference, and central difference forms are proportional to the orders of Δx, Δx, and $(\Delta x)^2$, respectively.

7.3   Fundamentals of FDTD

FDTD first introduced by Yee in 1966 [3] can be described as one of the general numerical method in electromagnetism. It originally lies in the preliminary resolution of the scattering problem in time domain. The algorithm is based on a central difference solution of Maxwell's equations with spatially staggered electric and magnetic fields that are solved alternately at each time step in a leapfrog algorithm. The harmonic solutions are obtained in a second step through a Fourier transformation. The FDTD algorithm implemented with PMLs works well for finding out the propagation constants of modes as a function of wavelength or normalized frequency for any arbitrary periodic boundary condition (PBC) structure that is uniform along the direction of propagation. A major advantage of the FDTD method, although it is time consuming and needs a large memory capacity, is that fairly accurate calculations are possible.

The FDTD formulation starts from the time-dependent differential form of Faraday's and Ampere's law or Maxwell's curl equations.

$$\frac{\partial \vec{B}}{\partial t} = -\nabla \times \vec{E}, \tag{7.5}$$

$$\frac{\partial \vec{D}}{\partial t} = -\nabla \times \vec{H} - \vec{J}, \tag{7.6}$$

where $\vec{J}$ is the current density. These equations in their general form can be solved either in a 1D, 2D, or 3D medium that has different properties, as presented in the following subsections.

7.3.1 1D Problem in Free Space

For the sake of clarity, we start from a simple situation: the materials are assumed to have isotropic, nondispersive, electric, and magnetic properties. Equations (7.5) and (7.6) can be redefined for free space as follows:

$$\frac{\partial \vec{E}}{\partial t} = -\frac{1}{\varepsilon_0} \nabla \times \vec{H}. \tag{7.7}$$

$$\frac{\partial \vec{H}}{\partial t} = -\frac{1}{\mu_0} \nabla \times \vec{E}. \tag{7.8}$$

Here, $\varepsilon_0 = 8.85 \times 10^{-12}$ F/m and $\mu_0 = 4\pi \times 10^7$ H/m.

As indicated by Eqs. (7.7) and (7.8), the field $\vec{E}$ and $\vec{H}$ components are vectors in 3D, but if we consider one propagation dimension, for example, along the z-axis, these equations can be expressed as a set of two coupled scalar equations in Cartesian coordinates as follows:

$$\frac{\partial \overrightarrow{E_x}}{\partial t} = -\frac{1}{\varepsilon_0} \frac{\partial \overrightarrow{H_y}}{\partial z}. \tag{7.9}$$

$$\frac{\partial \overrightarrow{H_y}}{\partial t} = -\frac{1}{\mu_0} \frac{\partial \overrightarrow{E_x}}{\partial z}. \tag{7.10}$$

The second step in the FDTD formulation is to approximate the derivatives with the "finite-difference" approximations as represented as follows:

$$\frac{E_x^{n+0.5} - E_x^{n-0.5}}{\Delta t} = -\frac{1}{\varepsilon_0} \frac{H_y^n(k+0.5) - H_y^n(k-0.5)}{\Delta z}.$$ (7.11)

$$\frac{H_y^{n+1}(k+0.5) - H_y^n(k+0.5)}{\Delta t} = -\frac{1}{\mu_0} \frac{E_x^{n+0.5}(k+1) - E_x^{n+0.5}(k)}{\Delta z}.$$ (7.12)

Here, the superscript n stands for a time $t = \Delta t n$ and k for the distance $z = \Delta z k$, where Δt and Δz are the time and spatial increments in time and 1D space, respectively.

The discretized equations can be rearranged to solve expressions for E_x and H_y as shown in the following:

$$E_x^{n+0.5}(k+0.5) = E_x^{n-0.5}(k) - \frac{\Delta t}{\Delta z \cdot \varepsilon_0}\left[H_y^n(k+0.5) - H_y^n(k-0.5)\right].$$ (7.13)

$$H_y^{n+1}(k+0.5) = H_y^n(k+0.5) - \frac{\Delta t}{\Delta z \cdot \mu_0}\left[E_x^{n+0.5}(k+1) - E_x^{n+0.5}(k)\right].$$ (7.14)

Equations (7.13) and (7.14) look very similar and the calculations are interleaved in both space and time. In this regard, the new value of E_x is calculated from the previous value of E_x and the most recent values of H_y. However, ε_0 and μ_0 differ by several orders of magnitude, therefore, E_x and H_y will also differ by several orders of magnitude. This can be circumvented by normalizing the electric field as shown in Eq. (7.15).

$$\tilde{E} = \sqrt{\frac{\varepsilon_0}{\mu_0}} E.$$ (7.15)

By substituting Eq. (7.15) into Eqs. (7.13) and (7.14), we can obtain the following equations:

$$\tilde{E}_x^{n+0.5}(k+0.5) = \tilde{E}_x^{n-0.5}(k) - \frac{\Delta t}{\Delta z \sqrt{\mu_0 \varepsilon_0}}\left[H_y^n(k+0.5) - H_y^n(k-0.5)\right].$$ (7.16)

$$H_y^{n+1}(k+0.5) = H_y^n(k+0.5) - \frac{\Delta t}{\Delta z \cdot \sqrt{\mu_0 \varepsilon_0}}\left[\tilde{E}_x^{n+0.5}(k+1) - \tilde{E}_x^{n+0.5}(k)\right].$$ (7.17)

It is revealed from these equations that $\tilde{E}$ and H will have the same order of magnitude.

7.3.2   1D Problem in a Lossless Medium

This section considers the case of a nonmagnetic medium that has a relative dielectric constant other than 1. That means that equations starting from (7.9) up to (7.17) must be modified to include the effect of ε_r as follows:

$$\frac{\partial \overrightarrow{E_x}}{\partial t} = -\frac{1}{\varepsilon_0 \varepsilon_r} \frac{\partial \overrightarrow{H_y}}{\partial z}. \tag{7.18}$$

$$\frac{E_x^{n+0.5} - E_x^{n-0.5}}{\Delta t} = -\frac{1}{\varepsilon_0 \varepsilon_r} \frac{H_y^n(k+0.5) - H_y^n(k-0.5)}{\Delta z}. \tag{7.19}$$

$$E_x^{n+0.5}(k+0.5) = E_x^{n-0.5}(k) - \frac{\Delta t}{\Delta z \cdot \varepsilon_0 \varepsilon_r} \left[H_y^n(k+0.5) - H_y^n(k-0.5) \right]. \tag{7.20}$$

$$\tilde{E} = \sqrt{\frac{\varepsilon_0 \varepsilon_r}{\mu_0}} E. \tag{7.21}$$

$$\tilde{E}_x^{n+0.5}(k+0.5) = \tilde{E}_x^{n-0.5}(k) - \frac{\Delta t}{\Delta z \sqrt{\mu_0 \varepsilon_0 \varepsilon_r}} \left[H_y^n(k+0.5) - H_y^n(k-0.5) \right]. \tag{7.22}$$

$$H_y^{n+1}(k+0.5) = H_y^n(k+0.5) - \frac{\Delta t}{\Delta z \sqrt{\mu_0 \varepsilon_0 \varepsilon_r}} \left[\tilde{E}_x^{n+0.5}(k+1) - \tilde{E}_x^{n+0.5}(k) \right]. \tag{7.23}$$

7.3.3   1D Problem in a Lossy Medium

To simulate the propagation in media that have conductivity, the general forms of Maxell's curl equations shown in Eq. (7.5) will be used after replacing the value of the current density J by $\sigma \vec{E}$ and dividing by the dielectric constant.

$$\frac{\partial \vec{E}}{\partial t} = -\frac{1}{\varepsilon_0 \varepsilon_r} \nabla \times \vec{H} - \frac{\sigma}{\varepsilon_0 \varepsilon_r} \vec{E}. \tag{7.24}$$

$$\frac{\partial \vec{H}}{\partial t} = -\frac{1}{\mu_0} \nabla \times \vec{E}. \tag{7.25}$$

Now, consider a simple 1D propagation along the z-axis. Therefore, Eq. (7.24) can be expressed as a scalar equation in Cartesian coordinates as shown in the following equation:

$$\frac{\partial E_x(t)}{\partial t} = -\frac{1}{\varepsilon_0 \varepsilon_r} \frac{\partial H_y(t)}{\partial z} - \frac{\sigma}{\varepsilon_0 \varepsilon_r} E_x(t). \tag{7.26}$$

Next, applying normalization using the variables of Eq. (7.15) gives

$$\frac{\partial \tilde{E}_x(t)}{\partial t} = -\frac{1}{\varepsilon_r \sqrt{\mu_0 \varepsilon_0}} \frac{\partial H_y(t)}{\partial z} - \frac{\sigma}{\varepsilon_0 \varepsilon_r} \tilde{E}_x(t), \tag{7.27}$$

$$\frac{\partial H_y(t)}{\partial t} = -\frac{1}{\sqrt{\mu_0 \varepsilon_0}} \frac{\partial \tilde{E}_x(t)}{\partial z}. \tag{7.28}$$

Finally, applying the FD approximations for both the temporal and spatial derivatives results in the following:

$$\frac{\tilde{E}_x^{n+0.5}(k) - \tilde{E}_x^{n-0.5}(k)}{\Delta t} = \frac{1}{\varepsilon_r \sqrt{\mu_0 \varepsilon_0}} \frac{H_y^n(k+0.5) - H_y^n(k-0.5)}{\Delta z} - \frac{\sigma}{\varepsilon_0 \varepsilon_r} \frac{\tilde{E}_x^{n+0.5}(k) + \tilde{E}_x^{n-0.5}(k)}{2} \tag{7.29}$$

$$\frac{H_y^{n+1}(k+0.5) - H_y^n(k+0.5)}{\Delta t} = -\frac{1}{\varepsilon_r \sqrt{\mu_0 \varepsilon_0}} \frac{\tilde{E}_x^{n+0.5}(k+1) - \tilde{E}_x^{n+0.5}(k)}{\Delta z} \tag{7.30}$$

The discretized equations can be rearranged to solve expressions for E_x and H_y as shown by Eqs. (7.31) and (7.32):

$$\tilde{E}_x^{n+0.5}(k) = \frac{\left[1 - \left(\Delta t \cdot \sigma / 2\varepsilon_0 \varepsilon_r\right)\right]}{\left[1 + \left(\Delta t \cdot \sigma / 2\varepsilon_0 \varepsilon_r\right)\right]} \tilde{E}_x^{n-0.5}(k) + \frac{\Delta t}{\Delta z \sqrt{\mu_0 \varepsilon_0}\left[\varepsilon_r + \frac{\Delta t \cdot \sigma}{2\varepsilon_0}\right]} \left[H_y^n(k+0.5) - H_y^n(k-0.5)\right] \tag{7.31}$$

$$H_y^{n+1}(k+0.5) = H_y^n(k+0.5) - \frac{\Delta t}{\Delta z \cdot \varepsilon_r \sqrt{\mu_0 \varepsilon_0}} \left[E_x^{n+0.5}(k+1) - E_x^{n+0.5}(k)\right] \tag{7.32}$$

7.3.4 2D Problem

In a 2D system, the fields can be decoupled into two transversely polarized modes, namely the TM$_z$ polarization (E_z, H_x, and H_y) and the TE$_z$ polarization (H_z, E_x, and E_y) as shown in Figure 7.1. Maxwell's equations are first projected onto the axes, before being discretized along x, y, and t. For example, the projection on the axes yields the following equations for the TM$_z$ polarization case:

$$\frac{\partial E_z}{\partial y} = -\mu_o \frac{\partial H_x}{\partial t}, \tag{7.33}$$

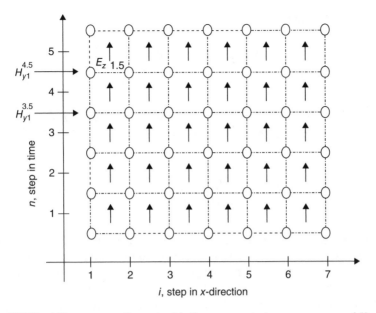

Figure 7.1 FDTD grid in one space dimension(x), E_z components shown as arrows and H_y components as squares. *Source*: Ref. [1]

$$\frac{\partial E_z}{\partial x} = \mu_o \frac{\partial H_y}{\partial t}, \tag{7.34}$$

$$\frac{\partial H_y}{\partial x} - \frac{\partial H_x}{\partial y} = \varepsilon_o \varepsilon_r \frac{\partial E_z}{\partial t}. \tag{7.35}$$

The following 2D FDTD time-stepping formulas constitute the discretization (in space and time) of Maxwell's equations on a discrete 2D mesh in a Cartesian x, y coordinate system for the TM$_z$ polarization case:

$$H_x\Big|_{i,j}^{n+\frac{1}{2}} = H_x\Big|_{i,j}^{n-\frac{1}{2}} - \frac{\Delta t}{\mu_{i,j}}\left[\frac{E_z\big|_{i,j+1}^{n} - E_z\big|_{i,j}^{n}}{\Delta y}\right], \tag{7.36}$$

$$H_y\Big|_{i,j}^{n+\frac{1}{2}} = H_y\Big|_{i,j}^{n-\frac{1}{2}} + \frac{\Delta t}{\mu_{i,j}}\left[\frac{E_z\big|_{i+1,j}^{n} - E_z\big|_{i,j}^{n}}{\Delta x}\right], \tag{7.37}$$

$$E_z\Big|_{i,j}^{n+1} = \frac{\varepsilon_{i,j} - \sigma_{i,j}\Delta t / 2}{\varepsilon_{i,j} + \sigma_{i,j}\Delta t / 2} E_z\Big|_{i,j}^{n}$$

$$+ \frac{\Delta t}{\varepsilon_{i,j} + \sigma_{i,j}\Delta t / 2}\left[\frac{H_y\big|_{i,j}^{n+1/2} - H_y\big|_{i-1,j}^{n+1/2}}{\Delta x} - \frac{H_x\big|_{i,j}^{n+1/2} - H_x\big|_{i,j-1}^{n+1/2}}{\Delta y}\right], \tag{7.38}$$

and for the TE$_z$ polarization case:

$$E_x\Big|_{i,j}^{n+\frac{1}{2}} = \frac{\varepsilon_{i,j} - \sigma_{i,j}\Delta t/2}{\varepsilon_{i,j} + \sigma_{i,j}\Delta t/2} E_x\Big|_{i,j}^{n} + \frac{\Delta t}{\varepsilon_{i,j} + \sigma_{i,j}\Delta t/2}\left[\frac{H_z\Big|_{i,j}^{n+1/2} - E_z\Big|_{i,j-1}^{n+1/2}}{\Delta y}\right], \tag{7.39}$$

$$E_y\Big|_{i,j}^{n+1} = \frac{\varepsilon_{i,j} - \sigma_{i,j}\Delta t/2}{\varepsilon_{i,j} + \sigma_{i,j}\Delta t/2} E_y\Big|_{i,j}^{n} - \frac{\Delta t}{\varepsilon_{i,j} + \sigma_{i,j}\Delta t/2}\left[\frac{H_z\Big|_{i,j}^{n+1/2} - E_z\Big|_{i-1,j}^{n+1/2}}{\Delta x}\right], \tag{7.40}$$

$$H_z\Big|_{i,j}^{n+1/2} = H_z\Big|_{i,j}^{n-1/2} - \frac{\Delta t}{\mu_{i,j}}\left[\frac{E_y\Big|_{i+1,j}^{n} - E_y\Big|_{i,j}^{n}}{\Delta x} - \frac{E_x\Big|_{i,j+1}^{n} - E_x\Big|_{i,j}^{n}}{\Delta y}\right]. \tag{7.41}$$

where the superscript n indicates the discrete time step, subscripts i and j denote the position of a grid point in the x- and y-directions, respectively. Δt is the time increment, and Δx and Δy are the space increments between two neighboring grid points along x- and y-directions, respectively. Additionally, ω is the angular frequency, μ, ε, and σ are the permeability, permittivity and the conductivity of the considered medium, respectively.

The performed discretization leads to an error of the second order in Δx or Δy. It is clear that in a 1D problem, the field components do not vary with y or z, then the update of the $E_z\Big|_{i}^{n+1}$ component would be as follows:

$$E_z\Big|_{i}^{n+1} = \frac{\varepsilon_i - \sigma_i\Delta t/2}{\varepsilon_i + \sigma_i\Delta t/2} E_z\Big|_{i}^{n} + \frac{\Delta t}{\varepsilon_i + \sigma_i\Delta t/2}\left[\frac{H_y\Big|_{i}^{n+1/2} - H_y\Big|_{i-1}^{n+1/2}}{\Delta x}\right]. \tag{7.42}$$

Figure 7.1 illustrates the locations of the field components in 1D space. From the figure, one can see that the fields do not interact with each other according to Maxwell's equations and the algorithm deals with scalars and does not understand that these are components of vectors.

7.3.5 3D Problem

Figure 7.2 shows the original FDTD paradigm described by Yee [3]. It is evident from the figure that the E and H fields are assumed interleaved around a cell whose origin is at the location i, j, and k. Every E field is located a half-cell width from the origin in the direction of its orientation. Also, every H field is offset by a half-cell in each direction except that of its orientation.

The general forms of Maxwell's curl equation represented by Eqs. (7.5) and (7.6) can be expressed as a set of six coupled scalar equations in Cartesian coordinates as shown in the following:

$$\frac{\partial H_x}{\partial t} = \frac{1}{\mu}\left[\frac{\partial E_y}{\partial z} - \frac{\partial E_z}{\partial y}\right]. \tag{7.43}$$

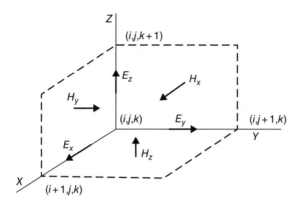

Figure 7.2   3D Yee cell

$$\frac{\partial H_y}{\partial t} = \frac{1}{\mu}\left[\frac{\partial E_z}{\partial x} - \frac{\partial E_x}{\partial z}\right]. \tag{7.44}$$

$$\frac{\partial H_z}{\partial t} = \frac{1}{\mu}\left[\frac{\partial E_x}{\partial y} - \frac{\partial E_y}{\partial x}\right]. \tag{7.45}$$

$$\frac{\partial E_x}{\partial t} = \frac{1}{\varepsilon}\left[\frac{\partial H_z}{\partial y} - \frac{\partial H_y}{\partial z} - \sigma E_x\right]. \tag{7.46}$$

$$\frac{\partial E_y}{\partial t} = \frac{1}{\varepsilon}\left[\frac{\partial H_x}{\partial z} - \frac{\partial H_z}{\partial x} - \sigma E_y\right]. \tag{7.47}$$

$$\frac{\partial E_z}{\partial t} = \frac{1}{\varepsilon}\left[\frac{\partial H_y}{\partial x} - \frac{\partial H_x}{\partial y} - \sigma E_z\right]. \tag{7.48}$$

Applying the difference approximation to equations starting from (7.43) up to (7.48) yields:

$$\frac{H_x^{n+1}\left(i, j+0.5, k+0.5\right) - H_x^{n-0.5}\left(i, j+0.5, k+0.5\right)}{\Delta t}$$
$$= -\frac{1}{\mu}\left[\frac{E_y^n\left(i, j+0.5, k+1\right) - E_y^n\left(i, j+0.5, k\right)}{\Delta z} - \frac{E_z^n\left(i, j+1, k+0.5\right) - E_z^n\left(i, j, k+0.5\right)}{\Delta y}\right]. \tag{7.49}$$

$$\frac{E_x^{n+1}\left(i+0.5, j, k\right) - E_x^n\left(i+0.5, j, k\right)}{\Delta t}$$
$$= -\frac{1}{\varepsilon}\left[\frac{H_y^{n+0.5}\left(i+0.5, j, k+1.5\right) - H_y^{n+0.5}\left(i+0.5, j, k+0.5\right)}{\Delta z}\right.$$
$$\left. -\frac{H_z^{n+0.5}\left(i+0.5, j+1.5, k\right) - H_z^{n+0.5}\left(i+0.5, j+0.5, k\right)}{\Delta y} + \sigma \cdot E_x^{n+0.5}\left(i+0.5, j, k\right)\right]. \tag{7.50}$$

Since,

$$E_x^{n+0.5}(i+0.5, j, k) = \frac{E_x^{n+1}(i+0.5, j, k) - E_x^n(i+0.5, j, k)}{\Delta t}. \tag{7.51}$$

Substituting Eq. (7.51) into Eq. (7.50) gives an explicit relation for E_x:

$$
\begin{aligned}
& E_x^{n+1}(i+0.5, j, k) \\
& = \left(\frac{1-(\sigma \cdot \Delta t / 2\varepsilon)}{1+(\sigma \cdot \Delta t / 2\varepsilon)}\right) E_x^n(i+0.5, j, k) \\
& + \left(\frac{\Delta t / \varepsilon}{1+(\sigma \Delta t / 2\varepsilon)}\right) \left(\begin{array}{c} \dfrac{H_z^{n+0.5}(i+0.5, j+1, k+0.5) - H_z^{n+0.5}(i+0.5, j, k+0.5)}{\Delta y} \\[3mm] -\dfrac{H_y^{n+0.5}(i+0.5, j, k+1.5) - H_y^{n+0.5}(i+0.5, j, k+0.5)}{\Delta z} \end{array}\right).
\end{aligned}
\tag{7.52a}
$$

A similar process can be applied to the other components to obtain the other updating equations. The other electric field updating equation sets can be expressed as follows:

$$
\begin{aligned}
& E_y^{n+1}(i, j+0.5, k) \\
& = \left(\frac{1-(\sigma \cdot \Delta t / 2\varepsilon)}{1+(\sigma \cdot \Delta t / 2\varepsilon)}\right) E_y^n(i, j+0.5, k) \\
& + \left(\frac{\Delta t / \varepsilon}{1+(\sigma \cdot \Delta t / 2\varepsilon)}\right) \left(\begin{array}{c} \dfrac{H_x^{n+0.5}(i, j+0.5, k+1.5) - H_x^{n+0.5}(i+0.5, j, k+0.5)}{\Delta z} \\[3mm] -\dfrac{H_z^{n+0.5}(i+1.5, j+0.5, k) - H_z^{n+0.5}(i+0.5, j+0.5, k)}{\Delta x} \end{array}\right).
\end{aligned}
\tag{7.52b}
$$

$$
\begin{aligned}
& E_z^{n+1}(i, j, k+0.5) \\
& = \left(\frac{1-(\sigma \cdot \Delta t / 2\varepsilon)}{1+(\sigma \cdot \Delta t / 2\varepsilon)}\right) E_z^n(i, j, k+0.5) \\
& + \left(\frac{\Delta t / \varepsilon}{1+(\sigma \cdot \Delta t / 2\varepsilon)}\right) \left(\begin{array}{c} \dfrac{H_y^{n+0.5}(i+1, j+0.5, k+0.5) - H_y^{n+0.5}(i, j+1, k+0.5)}{\Delta x} \\[3mm] -\dfrac{H_x^{n+0.5}(i, j+1.5, k+0.5) - H_x^{n+0.5}(i, j+0.5, k+0.5)}{\Delta y} \end{array}\right).
\end{aligned}
\tag{7.52c}
$$

The magnetic field updating equation sets are given by

$$
H_x^{n+0.5}\left(i,j+0.5,k+0.5\right)
$$
$$
= H_x^{n-0.5}\left(i,j+0.5,k+0.5\right) \tag{7.53a}
$$
$$
+\left(\frac{\Delta t}{\mu}\right)\left(\begin{array}{c} \dfrac{E_y^n\left(i,j+0.5,k+1\right)-E_y^n\left(i,j+0.5,k\right)}{\Delta z}- \\[4mm] \dfrac{E_z^n\left(i,j+1,k+0.5\right)-E_z^n\left(i,j,k+0.5\right)}{\Delta y} \end{array}\right).
$$

$$
H_y^{n+0.5}\left(i+0.5,j,k+0.5\right)
$$
$$
= H_y^{n-0.5}\left(i+0.5,j,k+0.5\right) \tag{7.53b}
$$
$$
+\left(\frac{\Delta t}{\mu}\right)\left(\begin{array}{c} \dfrac{E_z^n\left(i,j+0.5,k\right)-E_z^n\left(i+0.5,j,k\right)}{\Delta x}- \\[4mm] \dfrac{E_x^n\left(i+0.5,j,k+1\right)-E_x^n\left(i+0.5,j,k\right)}{\Delta z} \end{array}\right).
$$

$$
H_x^{n+0.5}\left(i,j+0.5,k+0.5\right)
$$
$$
= H_x^{n-0.5}\left(i,j+0.5,k+0.5\right) \tag{7.53c}
$$
$$
+\left(\frac{\Delta t}{\mu}\right)\left(\begin{array}{c} \dfrac{E_x^n\left(i+0.5,j+1,k\right)-E_x^n\left(i+0.5,j,k\right)}{\Delta y}- \\[4mm] \dfrac{E_y^n\left(i+1,j+0.5,k\right)-E_y^n\left(i,j+0.5,k\right)}{\Delta x} \end{array}\right).
$$

It can be revealed from these expressions that the new value of any electromagnetic field vector component at any point depends only on its previous value, the previous value of the adjacent points of the other field components.

7.4   Stability for FDTD

We cannot choose Δx, Δy, and Δt arbitrarily without stability problems. If the factor $\Delta t\,(\mu\,\Delta y)-1$ in Eq. (7.53a) would be too large, then H_x may receive too large part in each step in time. Therefore, H_x could grow toward infinity. For 2D FDTD, the stability limit for the orthogonal case is given by [3]:

$$
\Delta t \le \frac{1}{c\sqrt{\left(\dfrac{1}{\Delta x}\right)^2+\left(\dfrac{1}{\Delta y}\right)^2\left(\dfrac{1}{\Delta z}\right)^2}} \tag{7.54}
$$

where c is the propagation speed of the wave.

Most often, the above stability relation is written as $CFL < 1$, where CFL is the Courant–Friedrichs–Lewy number and is defined as follows:

$$CFL = c\Delta t \sqrt{\left(\frac{1}{\Delta x}\right)^2 + \left(\frac{1}{\Delta y}\right)^2 + \left(\frac{1}{\Delta z}\right)^2}. \tag{7.55}$$

7.5   Feeding Formulation

In practice, one often analyzes propagation of the Gaussian pulse or modulated Gaussian pulse. A Gaussian pulse centered at t_0 is defined as [12] follows:

$$g(t) = \exp\left[-\frac{(t - t_0)}{w^2}\right]. \tag{7.56}$$

Here, w is the width of the pulse in space. Further, the modulation signal is given by

$$m(t) = \sin\left[2\pi \frac{t}{T_0}\right], \tag{7.57}$$

where, T_0 is the period of the modulation signal $m(t)$. A continuous Gaussian pulse can be created by multiplying Eqs. (7.56) and (7.57)

$$g_m(t) = \sin\left[2\pi \frac{t}{T_0}\right] \exp\left[-\frac{(t - t_0)^2}{w^2}\right] \tag{7.58}$$

Equation (7.57) can be discretized as follows:

$$g_m(n) = \sin\left[2\pi \Delta \frac{n}{T_0}\right] \exp\left[-\frac{(n\Delta t - t_0)^2}{w^2}\right] \tag{7.59}$$

7.6   Absorbing Boundary Conditions

A basic consideration with the FD approach in electromagnetic wave interaction problems is that many geometries of interest are defined in "open" regions. That spatial domain of the computed field is unbounded in one or more coordinate directions. Clearly, no computer can store an infinite amount of data, and therefore the field computation domain must be limited in size. Therefore, absorbing boundary conditions (ABCs) are necessary to completely absorb the waves and keep outgoing electromagnetic fields from being reflected back. In order to compute the E field, the surrounding H values must be known. However, at the cutting edge we will not have the value to one side, but we know that there are no sources outside the simulation space and the fields at the edge must be propagating outward. In the following subsections, Mur's ABC and the PML ABC that absorb the wave in the outer boundary of the computational lattice will be briefly discussed.

7.6.1 Mur's ABCs

FD schemes for the first- and second-order ABCs are introduced by Mur [13]. In this book [13], Mur's first-order ABC has been adopted only for a 1D FDTD code. This scheme assumes that the pulse lines will be normally incident to the mesh walls, where the tangential fields on the outer boundaries will obey a 1D wave equation in the direction normal to the mesh wall. For x normal wall, the 1D wave equation may be written for low x (the edge at $x=0$) with discretized form as follows:

$$E_{y|i=1}^{n+1} = E_{y|i=2}^{n+1} - \frac{c\Delta t - \Delta x}{c\Delta t + \Delta x}\left[E_{y|i=1}^{n} - E_{y|i=2}^{n}\right].$$ (7.60)

Here, $E_{y|i=1}$ represents the tangential electric field component on the mesh wall and $E_{y|i=2}$ represents the tangential electric field components on a node inside the mesh wall. The above expression is known Mur's first-order ABCs. A similar relation can be obtained for Mur's second-order radiation boundary conditions [2].

7.6.2 Perfect Matched Layer

The PML ABC, first proposed and developed by Berenger [14] in 1994, is one of the most flexible, efficient, and powerful tools in the FDTD technique. Berenger's method is basically the truncation of the computational domain by a layer that absorbs the impinging plane waves with no reflection, irrespective of their frequency and angle of incidence. The innovation of Berenger's PML is that, the plane waves of arbitrary incidence, polarization, and frequency are matched at the boundary. PML is a dispersive medium with intrinsic wave impedance and phase velocity equal to the wave impedance of a vacuum and vacuum phase velocity, respectively. The work of Berenger was followed by a more general form called uniaxial PML (UPML) [15, 16]. One of the main disadvantages of PML is the need for additional memory requirements due to the process of modifying electric and magnetic fields through the PML.

To easily realize the PML in 3D, recall that any EM wave can be treated as a transverse magnetic (TM) wave plus transverse electric (TE) wave. Next, the complete 3D PML formulation can be easily realized [17]. Based on the 2D conditions, the directional conductivities $\sigma_x(x)$ and $\sigma_y(y)$ exist along only the lower and upper regions of PML along the x-axis and y-axis, respectively as follows:

$$j\omega\left(1+\frac{\sigma(x)}{j\omega\varepsilon_{o}}\right)\left(1+\frac{\sigma(y)}{j\omega\varepsilon_{o}}\right)D_{z} = C_{o}\left(\frac{\partial H_{y}}{\partial x} - \frac{\partial H_{x}}{\partial y}\right).$$ (7.61)

$$j\omega\left(1+\frac{\sigma(x)}{j\omega\varepsilon_{o}}\right)^{-1}\left(1+\frac{\sigma(y)}{j\omega\varepsilon_{o}}\right)B_{x} = -C_{o}\left(\frac{\partial E_{z}}{\partial y}\right).$$ (7.62)

$$j\omega\left(1+\frac{\sigma(x)}{j\omega\varepsilon_o}\right)\left(1+\frac{\sigma(y)}{j\omega\varepsilon_o}\right)^{-1}B_y = -C_o\left(\frac{\partial E_z}{\partial x}\right). \qquad (7.63)$$

Applying the difference approximation to Eqs. (7.61), (7.62), and (7.63) yields the following solutions:

$$D_z\big|_{i,j}^{n+1} = gi3(i)\cdot gi3(j)\cdot D_z\big|_{i,j}^{n} + gi2(i)\cdot gi2(j)\cdot c_0 \cdot \Delta t$$
$$\times\left[\frac{H_y\big|_{i+0.5,j}^{n+0.5} - H_y\big|_{i-0.5,j}^{n+0.5}}{\Delta x} - \frac{H_x\big|_{i,j+0.5}^{n+0.5} - H_x\big|_{i,j-0.5}^{n+0.5}}{\Delta y}\right]. \qquad (7.64)$$

$$B_x\big|_{i,j+0.5}^{n+0.5} = fi3(j+0.5)\cdot B_x\big|_{i,j+0.5}^{n-0.5} + fi2(j+0.5)$$
$$\times c_0\Delta t\times\left[-\left(\frac{E_z\big|_{i,j+1}^{n} - E_z\big|_{i,j}^{n}}{\Delta y}\right) + fi1(j)\times\sum_{i=1}^{n}-\left(\frac{E_z\big|_{i,j+1}^{n} - E_z\big|_{i,j}^{n}}{\Delta y}\right)\right]. \qquad (7.65)$$

$$B_y\big|_{i+0.5,j}^{n+0.5} = fi3(i+0.5)\cdot B_y\big|_{i+0.5,j}^{n-0.5} + fi2(i+0.5)$$
$$\times c_0\Delta t\times\left[-\left(\frac{E_z\big|_{i+1,j}^{n} - E_z\big|_{i,j}^{n}}{\Delta x}\right) + fj1(j)\times\sum_{i=1}^{n}-\left(\frac{E_z\big|_{i+0.5,j}^{n} - E_z\big|_{i,j}^{n}}{\Delta x}\right)\right]. \qquad (7.66)$$

Here, c_0 is the speed of light in vacuum and the definitions of $gi2(i)$, $gi3(i)$, $gj2(j)$, $gj3(j)$, $fi1(i)$, $fi2(i)$, $fi3(i)$, $fj1(j)$, $fj2(j)$, and $fj3(j)$ are defined as follows:

$$gi3(i) = \left(\frac{2\varepsilon_0 - \sigma_x(i)\Delta t}{2\varepsilon_0 + \sigma_x(i)\Delta t}\right). \qquad (7.67)$$

$$gj3(j) = \left(\frac{2\varepsilon_0 - \sigma_y(j)\Delta t}{2\varepsilon_0 + \sigma_y(j)\Delta t}\right). \qquad (7.68)$$

$$gi2(i) = \left(\frac{2\varepsilon_0}{2\varepsilon_0 + \sigma_x(i)\Delta t}\right). \qquad (7.69)$$

$$gj2(j) = \left(\frac{2\varepsilon_0}{2\varepsilon_0 + \sigma_y(j)\Delta t}\right)fi3(i+0.5) = \left(\frac{2\varepsilon_0 - \sigma_x(i+1)\Delta t}{2\varepsilon_0 + \sigma_x(i+0.5)\Delta t}\right). \qquad (7.70)$$

$$fj3(j+0.5) = \left(\frac{2\varepsilon_0 - \sigma_y(j+0.5)\Delta t}{2\varepsilon_0 + \sigma_y(j+0.5)\Delta t}\right). \qquad (7.71)$$

$$fi2(i+0.5) = \left(\frac{2\varepsilon_0}{2\varepsilon_0 + \sigma_x(i+0.5)\Delta t} \right). \tag{7.72}$$

$$fj2(j+0.5) = \left(\frac{2\varepsilon_0}{2\varepsilon_0 + \sigma_y(j+0.5)\Delta t} \right). \tag{7.73}$$

$$fj1(i) = \left(\frac{\sigma_x(i)\Delta t}{2\varepsilon_0} \right). \tag{7.74}$$

$$fj1(j) = \left(\frac{\sigma_y(j)\Delta t}{2\varepsilon_0} \right). \tag{7.75}$$

The formulation of Maxwell's equations for the TE mode can be given as follows:

$$j\omega \left(1 + \frac{\sigma(x)}{j\omega\varepsilon_o} \right) \left(1 + \frac{\sigma(y)}{j\omega\varepsilon_o} \right) B_z = -C_o \left(\frac{\partial E_x}{\partial y} - \frac{\partial E_y}{\partial x} \right). \tag{7.76}$$

$$j\omega \left(1 + \frac{\sigma(x)}{j\omega\varepsilon_o} \right)^{-1} \left(1 + \frac{\sigma(y)}{j\omega\varepsilon_o} \right) D_x = C_o \left(\frac{\partial H_z}{\partial y} \right). \tag{7.77}$$

$$j\omega \left(1 + \frac{\sigma(x)}{j\omega\varepsilon_o} \right) \left(1 + \frac{\sigma(y)}{j\omega\varepsilon_o} \right)^{-1} D_y = -C_o \left(\frac{\partial H_z}{\partial x} \right). \tag{7.78}$$

The difference approximations of the above equations are given by

$$B_z\big|_{i+0.5,j+0.5}^{n+0.5} = fi3(j+0.5) \cdot fj3(j+0.5) \cdot B_z\big|_{i+1,j+0.5}^{n-0.5} + fi2(i+0.5)$$

$$\times fj2(i+0.5) \times c_0\Delta t \times \left[\left(\frac{E_y\big|_{i+1,j+0.5}^n - E_y\big|_{i,j+0.5}^n}{\Delta x} \right) \left(\frac{E_x\big|_{i+0.5,j+1}^n - E_x\big|_{i+0.5,j}^n}{\Delta y} \right) \right] \tag{7.79}$$

$$D_x\big|_{i+0.5,j}^{n+1} = gj3(j) \cdot D_x\big|_{i+0.5,j}^n + gi2(j) \cdot c_0 \cdot \Delta t$$

$$\times \left[\frac{H_z\big|_{i+0.5,j+0.5}^{n+0.5} - H_z\big|_{i+0.5,j-0.5}^{n+0.5}}{\Delta y} + gi1(i+0.5) \times \sum_{i=1}^n \frac{H_z\big|_{i+0.5,j+0.5}^{n+0.5} - H_z\big|_{i+0.5,j-0.5}^{n+0.5}}{\Delta y} \right] \tag{7.80}$$

$$D_y\big|_{i,j+0.5}^{n+1} = gj3(j) \cdot D_y\big|_{i,j+0.5}^n + gi2(j) \cdot c_0 \cdot \Delta t$$

$$\times \left[\frac{H_z\big|_{i+0.5,j+0.5}^{n+0.5} - H_z\big|_{i-0.5,j+0.5}^{n+0.5}}{\Delta y} + gi1(i+0.5) \times \sum_{i=1}^n - \left(\frac{H_z\big|_{i+0.5,j+0.5}^{n+0.5} - H_z\big|_{i-0.5,j+0.5}^{n+0.5}}{\Delta x} \right) \right] \tag{7.81}$$

7.7   1D FDTD Sample Code

In this section, a MATLAB-based computer program is written to simulate the wave propagation in a 1D lossless nonmagnetic finite slab that is surrounded by air using perfect electric conductor (PEC) boundary conditions. The program is composed of three successive sections: source simulation, structure simulation, and propagation simulation. The first section has been written to define the source, the second section to define the geometrical parameters as well as the electric and magnetic properties of the structure, and the third section to simulate the field updating equations using the 1D FDTD technique. The code of each section will be explained in detail in the following subsections.

7.7.1   Source Simulation

In this subsection, the following code I has been written to define and visualize the source at the operating wavelength.

Code I

```
clear;
clc;
%[1]Defining the operating wavelength
lamdo=0.5e-6;                % the operating wavelength
c=3e+8;                      % the speed of light in free space
fo=c/lamdo;                  % the operating frequency
t=0:0.1e-15:40e-15;          % The simulation time
N=length(t);                 %length of time vector
%[2]Defining the source
tc=4.2e-15;
sigm=1.8e-15;
%Sinusoidal source
ssource=sin(2*pi*fo.*t);
%Gaussianl source
gsource=exp(-(t-tc).^2/(2*sigm^2));
%Modulated Gaussian
mgsource=sin(2*pi*fo.*t).*exp(-(t-tc).^2./(2*sigm^2));
%[3] Plotting the source
figure(1)
%Plotting the sinusoidal source
subplot221
plot(t.*1e+15,ssource)
xlabel('Time (fs)')
ylabel('Ampilitude')
%Plotting the Gaussian source
subplot222
plot(t.*1e+15,gsource)
xlabel('Time (fs)')
```

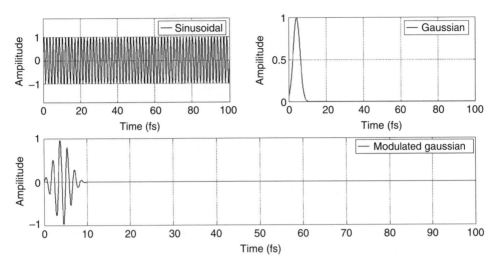

Figure 7.3 Source simulation: (a) sinusoidal, (b) Gaussian, and (c) modulated Gaussian

```
ylabel('Ampilitude')
%Plotting the Modulated Gaussian source
subplot212
plot(t.*1e+15,mgsource)
xlabel('Time (fs)')
ylabel('Ampilitude')
```

As can be seen, code I is divided into three parts. The first part is used to define the operating wavelength and hence the operating frequency. Further, the start, end, and step values of the simulation time are presented. The equations of the source are defined in the second part. Finally, the steps of the third part are added to plot the time-varying source. Figure 7.3 shows the sinusoidal, Gaussian, and modulated Gaussian signals for time duration of about 100 fs.

7.7.2   Structure Simulation

In this subsection, code II has been written to define the 1D structure. Code II can be also divided into two parts. The first part is used to define the geometries as well as the electric and magnetic properties of nonmagnetic 1D slab of length of 12.5 μm and relative permittivity of 2. In addition, the discretization of the structure is determined where the space discretized step Δz should be less than or equal to $\lambda/10$ and the time discretized step Δt should be equal to $\Delta z/c$ [1], where λ and c are defined as the operating wavelength and light speed in the free space, respectively. The code can help visualize the variation of the permittivity along the considered structure. Figure 7.4 shows the plot of distribution of the relative permittivity along the considered 1D structure. Part 2 of the code is written to define empty matrices for the unknown magnetic and electric fields. The length of the

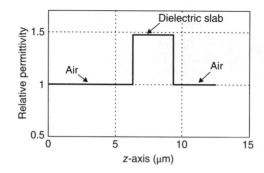

Figure 7.4   Structure simulation: distribution of the relative permittivity along the 1D structure

electric field vector E_x is greater than the structure length by only one cell to define the virtual boundary cell.

Code II

```
% [1]Defining the Structure
dz = lamdo/20;        % Space discretized step
dt = dz/c;            % Time discretized step
z = 0:dz:12.5e-6;     % The length of the 1-D slab
m = length(z);
muo = 4*pi*1e-7;
mu = muo.*ones(1,m);
epr = ones(1,m);
epr(m/2 + 1:3*m/4) = 1.48;
epo = 8.85418e-12;
ep = epo.*epr;
figure(2)
plot(1e + 6.*z,epr)
xlabel('z-axis (\ mico m)')
ylabel('Relative Permittivity')
%[2] Defining Empty matrices for magnetic & Electric fields
Hy = zeros(1,m);
Ex = zeros(1,m+1);
```

7.7.3   Propagation Simulation

In the final section of the presented 1D code, code III has been written to calculate and visualize the propagating fields along the considered structure defined previously in Section 7.7.2. Here, the code is based on placing the Gauss modulated source at the beginning of the structure (at cell 1) and then varying its amplitude versus the time. During the simulation time, the magnetic and electric fields are recorded and updated at each time step. This simulation exhibits perfect electric conductor boundary conditions, where the electric field at the virtual cell bounding the structure is defined by zero. The final steps of code III are added to visualize

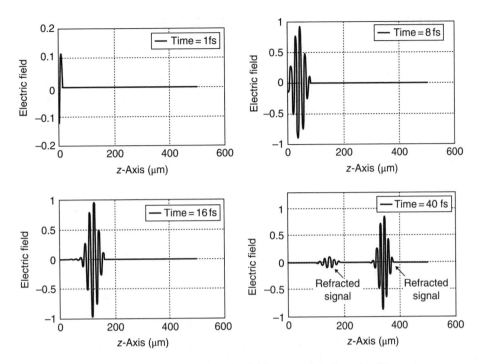

Figure 7.5   Four time snapshots for the field propagation along the 1D structure

the propagation of the electromagnetic fields along the considered structure. Figure 7.5 shows four snapshots for the propagating electric field along the considered 1D structure at different simulation times.

Code III

```
% Propagation
for n=1:N
% Defining the time varying source at cell 1
Ex(1)=mgsource(n);
% Updating magnetic and electric fields
Hy=Hy-dt./mu.*diff(Ex)./dz;
    Ex(2:m)=Ex(2:m)-dt./ep(2:m).*diff(Hy)./dz;
% Defining the Electric field at the PEC boundary
Ex(m+1)=0;
% Visualizing the Electric field
dd=plot(Ex(1:m));
xlabel('z-axis')
ylabel('Electric field')
p(n)=getframe;
end
movie(p,1)
```

7.8   FDTD Formulation for Anisotropic Materials

Certain optical structures such as liquid crystals, fibers, and ferrites exhibit anisotropic behavior. The term "anisotropic" indicates that the dielectric constant is different along the different crystal directions with respect to the crystal axes. Although the anisotropic property is sometimes a disadvantage, it can be used for controlling the optical properties of the studied structure. This section provides the formulations and special treatments of the FDTD method for analyzing such structures with arbitrary anisotropic permittivity tensors.

The general form of Maxwell's equations for anisotropic materials is as follows

$$\frac{\partial \vec{D}}{\partial t} = \nabla \times \vec{H}. \tag{7.82}$$

$$\vec{D}(\omega) = [\varepsilon] \cdot \vec{E}(\omega) \quad \text{or} \quad \vec{E}(\omega) = [\varepsilon]^{-1} \vec{D}. \tag{7.83}$$

Here,

$$[\varepsilon] = \begin{bmatrix} \varepsilon_{xx} & \varepsilon_{xy} & \varepsilon_{xz} \\ \varepsilon_{yx} & \varepsilon_{yy} & \varepsilon_{yz} \\ \varepsilon_{zx} & \varepsilon_{zy} & \varepsilon_{zz} \end{bmatrix}, \quad [\varepsilon]^{-1} = \frac{1}{|\varepsilon|} \begin{bmatrix} C_{xx} & C_{xy} & C_{xz} \\ C_{yx} & C_{yy} & C_{yz} \\ C_{zx} & C_{zy} & C_{zz} \end{bmatrix} \tag{7.84a}$$

and

$$|\varepsilon| = \varepsilon_0 \left(\varepsilon_{xx}\varepsilon_{yy}\varepsilon_{zz} + \varepsilon_{xy}\varepsilon_{yz}\varepsilon_{zx} + \varepsilon_{xz}\varepsilon_{yx}\varepsilon_{zy} - \varepsilon_{xz}\varepsilon_{yy}\varepsilon_{zx} - \varepsilon_{xy}\varepsilon_{yz}\varepsilon_{zz} - \varepsilon_{xx}\varepsilon_{yz}\varepsilon_{zy} \right). \tag{7.84b}$$

$$\frac{\partial \vec{B}}{\partial t} = \nabla \times \vec{E}. \tag{7.85}$$

$$\vec{B}(\omega) = [\mu] \cdot \vec{H}(\omega) \quad \text{or} \quad \vec{H}(\omega) = [\mu]^{-1} \vec{B} \tag{7.86}$$

$$[\mu] = \begin{bmatrix} \mu_{xx} & \mu_{xy} & \mu_{xz} \\ \mu_{yx} & \mu_{yy} & \mu_{yz} \\ \mu_{zx} & \mu_{zy} & \mu_{zz} \end{bmatrix}, \quad [\mu]^{-1} = \frac{1}{|\mu|} \begin{bmatrix} d_{xx} & d_{xy} & d_{xz} \\ d_{yx} & d_{yy} & d_{yz} \\ d_{zx} & d_{zy} & d_{zz} \end{bmatrix} \tag{7.87}$$

and

$$|\mu| = \mu_0 \left(\mu_{xx}\mu_{yy}\mu_{zz} + \mu_{xy}\mu_{yz}\mu_{zx} + \mu_{xz}\mu_{yz}\mu_{zx} + \mu_{xz}\mu_{yz}\mu_{zy} \right.$$
$$\left. - \mu_{xz}\mu_{yz}\mu_{zy} - \mu_{xz}\mu_{yy}\mu_{zx} - \mu_{xy}\mu_{yz}\mu_{zz} - \mu_{xx}\mu_{yz}\mu_{zy} \right) \tag{7.88}$$

The relationships between the electric field and the electric flux are as follows:

$$E_x\Big|_{i+0.5,j,k}^{n+1} = \frac{C_{xx}}{|\varepsilon|} D_x\Big|_{i+0.5,j,k}^{n+1} + \frac{C_{xy}}{|\varepsilon|} D_y\Big|_{i+0.5,j,k}^{n+1} + \frac{C_{xz}}{|\varepsilon|} D_z\Big|_{i+0.5,j,k}^{n+1}. \tag{7.89}$$

$$E_y\Big|_{i,j+0.5,k}^{n+1} = \frac{C_{yx}}{|\varepsilon|} D_x\Big|_{i,j+0.5,k}^{n+1} + \frac{C_{yy}}{|\varepsilon|} D_y\Big|_{i,j+0.5,k}^{n+1} + \frac{C_{yz}}{|\varepsilon|} D_z\Big|_{i,j+0.5,k}^{n+1}. \tag{7.90}$$

$$E_z\Big|_{i,j,k+0.5}^{n+1} = \frac{C_{xz}}{|\varepsilon|} D_x\Big|_{i,j,k+0.5}^{n+1} + \frac{C_{zy}}{|\varepsilon|} D_y\Big|_{i,j,k+0.5}^{n+1} + \frac{C_{zz}}{|\varepsilon|} D_z\Big|_{i,j,k+0.5}^{n+1}. \tag{7.91}$$

To obtain the expressions for D_x, D_y, and D_z, the central FD approximation has been applied as presented in the following:

$$D_y\Big|_{i+0.5,j,k} = 0.25\left(D_y\Big|_{i,j-0.5,k} + D_y\Big|_{i,j+0.5,k} + D_y\Big|_{i+1,j-0.5,k} + D_y\Big|_{i+1,j+0.5,k}\right). \tag{7.92}$$

$$D_z\Big|_{i+0.5,j,k} = 0.25\left(D_z\Big|_{i,j,k-0.5} + D_z\Big|_{i,j+0.5,k} + D_z\Big|_{i+1,j,k-0.5} + D_z\Big|_{i+1,j,k+0.5}\right). \tag{7.93}$$

$$D_x\Big|_{i,j+0.5,k} = 0.25\left(D_x\Big|_{i-0.5,j,k} + D_x\Big|_{i-0.5,j+1,k} + D_x\Big|_{i+0.5,j+1,k} + D_x\Big|_{i+0.5,j,k}\right). \tag{7.94}$$

$$D_z\Big|_{i,j+0.5,k} = 0.25\left(D_z\Big|_{i,j+1,k+0.5} + D_z\Big|_{i,j,k+0.5} + D_z\Big|_{i,j+1,k-0.5} + D_z\Big|_{i,j,k+0.5}\right). \tag{7.95}$$

$$D_x\Big|_{i,j,k+0.5} = 0.25\left(D_x\Big|_{i+0.5,j,k} + D_x\Big|_{i-0.5,j+1,k} + D_x\Big|_{i+0.5,j+1,k+1} + D_x\Big|_{i-0.5,j,k+1}\right). \tag{7.96}$$

$$D_y\Big|_{i,j,k+0.5} = 0.25\left(D_y\Big|_{i,j+0.5,k} + D_y\Big|_{i,j-0.5,k} + D_y\Big|_{i,j+0.5,k+!} + D_y\Big|_{i,j-0.5,k+1}\right). \tag{7.97}$$

The relationships between the magnetic field and the magnetic flux are as follows:

$$H_x\Big|_{i,j+0.5,k+0.5}^{n+1.5} = \frac{d_{xx}}{|\mu|} B_x\Big|_{i,j+0.5,k+0.5}^{n+1.5} + \frac{d_{xy}}{|\mu|} D_y\Big|_{i,j+0.5,k+0.5}^{n+1.5} + \frac{d_{xz}}{|\mu|} B_z\Big|_{i,j+0.5,k+0.5}^{n+1.5}. \tag{7.98}$$

$$H_y\Big|_{i+0.5,j,k+0.5}^{n+1.5} = \frac{d_{yx}}{|\mu|} B_x\Big|_{i+0.5,j,k+0.5}^{n+1.5} + \frac{d_{yy}}{|\mu|} D_y\Big|_{i+0.5,j,k+0.5}^{n+1.5} + \frac{d_{yz}}{|\mu|} B_z\Big|_{i+0.5,j,k+0.5}^{n+1.5}. \tag{7.99}$$

$$H_z\Big|_{i+0.5,j+0.5,k}^{n+1.5} = \frac{d_{xz}}{|\mu|} B_x\Big|_{i+0.5,j+0.5,k}^{n+1.5} + \frac{d_{zy}}{|\varepsilon\mu|} D_y\Big|_{i+0.5,j+0.5,k}^{n+1.5} + \frac{d_{zz}}{|\varepsilon\mu|} B_z\Big|_{i+0.5,j+0.5,k}^{n+1.5}. \tag{7.100}$$

Similarly, to obtain the expressions for B_x, B_y, and B_z, the central FD approximation has been applied as presented in the following:

$$B_y\big|_{i,j+0.5,k+0.5} = 0.25\left(B_y\big|_{i+0.5,j+1,k+0.5} + B_y\big|_{i+0.5,j,k+0.5} + B_y\big|_{i-0.5,j,k+0.5} + B_y\big|_{i-0.5,j,k+0.5}\right). \quad (7.101)$$

$$B_z\big|_{i,j+0.5,k+0.5} = 0.25\left(B_z\big|_{i+0.5,j+0.5,k+1} + B_z\big|_{i+0.5,j+0.5,k} + B_z\big|_{i-0.5,j+0.5,k+1} + B_z\big|_{i-0.5,j+0.5,k}\right). \quad (7.102)$$

$$B_x\big|_{i+0.5,j,k+0.5} = 0.25\left(B_x\big|_{i+1,j+0.5,k+0.5} + B_x\big|_{i+1,j-0.5,k+0.5} + B_x\big|_{i-0.5,j+1,k+0.5} + B_x\big|_{i,j-0.5,k+0.5}\right). \quad (7.103)$$

$$B_z\big|_{i+0.5,j,k+0.5} = 0.25\left(B_z\big|_{i+0.5,j+0.5,k+1} + B_z\big|_{i+1,j-0.5,k+0.5} + B_z\big|_{i+0.5,j-0.5,k+1} + B_z\big|_{i+0.5,j-0.5,k}\right). \quad (7.104)$$

$$B_x\big|_{i+0.5,j+0.5,k} = 0.25\left(B_x\big|_{i+1,j+0.5,k+0.5} + B_x\big|_{i,j+0.5,k+0.5} + B_x\big|_{i+1,j+0.5,k-0.5} + B_x\big|_{i,j+0.5,k-0.5}\right). \quad (7.105)$$

$$B_y\big|_{i+0.5,j+0.5,k} = 0.25\left(B_y\big|_{i+0.5,j+1,k+0.5} + B_y\big|_{i+0.5,j+1,k-0.5} + B_y\big|_{i+0.5,j,k+0.5} + B_y\big|_{i+0.5,j,k-0.5}\right). \quad (7.106)$$

7.9 Summary

In this chapter, the basic principles and formulations of the full-vectorial FDTD method are discussed in detail. The formulations of 1D FDTD for free space, lossless, and lossy media are first introduced followed by a 1D FDTD sample code. Next, the formulations of 2D and 3D FDTD for isotropic and anisotropic materials are presented. Further, the numerical stability and the ABCs of the FDTD method are also presented.

References

[1] Taflove, A. (1998) *Advances in Computational Electrodynamics: The Finite Difference Time Domain Method*, Artech House, Boston.
[2] Taflove, A. and Susan, C. (2000) *Hangness, Computational Electrodynamics: The finite Difference Time Domain Method*, Artech House, Boston.
[3] Yee, K.S. (1966) Numerical solution of initial boundary value problem involving Maxwell's equations in isotropics media. *IEEE Trans. Antennas Propag.*, **AP-14**, 302–307.
[4] Choi, D.H. and Hoefer, W.J.R. (1986) The finite difference time domain method and its application to eigenvalue problem. *IEEE Trans. Microw. Theory Tech.*, **34**, 1464–1470.
[5] Schneider, J. and Hudson, S. (1993) The finite difference time domain method applied to anisotropic material. *IEEE Trans. Antennas Propag.*, **41**, 994–999.
[6] Hunsberger, F. and Luebbers, R.J. (1992) Finite difference time domain analysis of gyrotropic media. I: magnetized plasma. *IEEE Trans. Antennas Propag.*, **40**, 1489–1495.
[7] Zhao, A. (1999) An efficient FDTD algorithm for the analysis of microstrip patch antennas printed on a general anisotropic dielectric substrate. *IEEE Trans. Microw. Theory Tech.*, **47** (7), 1142–1146.
[8] Moss, C.D. and Teixeira, F.L. (2002) Analysis and compensation of numerical dispersion in the FDTD method for layered anisotropic media. *IEEE Trans. Antennas Propag.*, **50**, 1174–1184.
[9] Akyurtlu, A. and Werner, D.H. (July 1999) Modeling of transverse propagation through a uniaxial bianisotropic medium using the finite difference time domain technique. *IEEE Trans. Antennas Propag.*, **47** (7), 1142–1146.

[10] Mosallaei, H. and Sarabandi, K. (2004) Magneto-dielectrics in electromagnetics: concept and applications. *IEEE Trans. Antennas Propag.*, **52**, 1558–1567.

[11] Sadiku, M.N.O. (1992) *Numerical Techniques in Electromagnetics*, CRC Press, Boca Raton.

[12] Umashankar, K.R. (1999) Finite-difference time domain method, in *Time Domain Electromagnetics* (ed S.M. Rao), Academic Press, San Diego, pp. 151–235.

[13] Wartak, M.S. (2013) *Computational Photonics: An Introduction with MATLAB*, Cambridge University Press, Cambridge.

[14] Berenger, J.P. (1996) A perfectly matched layer for the absorption of electromagnetic waves. *J. Comput. Phys.*, **6**, 97–99.

[15] Gedney, S.D. (1996) An anisotropic perfectly matched layer absorbing media for the truncation of FDTD lattices. *IEEE Trans. Antennas Propag.*, **44**, 1630–1639.

[16] Gedney, S. (2003) *Computational Electromagnetic: The Finite Difference Time Domain*, Lecture notes, Fall 2003. Artech House, Boston.

[17] Balanis, C. (2001) *Advanced Engineering Electromagnetic Theory*, John Wiley & Sons, Inc., New York.

Part III

Applications of LC Devices

8

Polarization Rotator Liquid Crystal Fiber

8.1 Introduction

Only in recent years have industries that manufacture and use photonic devices become aware of the unexpected and undesirable polarization rotation, and consequent polarization crosstalk (CT) that can occur in optoelectronic systems. This effect is a cause of particular concern as, in the presence of discontinuities in the propagation direction, the optical modes can exchange power if they are nearly phase matched and have a significant overlap. In a complex semiconductor optoelectronic system, in particular hybrid modes can experience an exchange of power between the polarization states due to the presence of junctions, tapers, bends, or other discontinuities. As a result, an important issue is to understand the physical process in such polarization conversions and to consider an appropriate design approach for its minimization.

Fortunately, knowledge of the origin of such polarization CT can be used in the design of compact low-loss polarization rotators (PRs). PRs [1, 2] are key components in optoelectronic integrated circuits for coherent optical system applications, employed particularly to control polarization states. They may also be used in balanced polarization diversity heterodyne receivers to provide a constant 45° operation. Compact low-loss PRs may be used in the design of future generations of monolithically integrated optical isolators.

In this chapter, highly tunable PRs based on photonic crystal fiber (PCF) infiltrated with nematic liquid crystal (NLC) [3, 4] are introduced and analyzed. First, an overview of the previously reported PRs is presented, followed by practical applications of PRs. Second, the operation principles of PRs and numerical simulation strategy and analysis of the NLC photonic crystal fiber (NLC-PCF) PRs are presented. Last, the fabrication aspects of the proposed PRs are introduced. The analysis is made by the full-vectorial finite-difference method (FVFDM) [5], and FVFD beam propagation method (FVFD-BPM) [6].

Computational Liquid Crystal Photonics: Fundamentals, Modelling and Applications, First Edition.
Salah Obayya, Mohamed Farhat O. Hameed and Nihal F.F. Areed.
© 2016 John Wiley & Sons, Ltd. Published 2016 by John Wiley & Sons, Ltd.

8.2   Overview of PRs

The PR is a vital component in communication systems and can be used to control the polarization states in communication systems such as polarization modulators [1] and polarization switches [2]. Initially, a PR based on multi-section structures with periodically alternating asymmetric loading [7] was proposed. This approach suffers from its longer device length (several millimeters) and large transition losses at the interface between alternating sections (several decibels). However, Obayya *et al.* [8] reported that by careful adjustment of the waveguide width and/or the waveguide materials, one can obtain complete polarization conversion at a moderate device length around 0.74 mm, and with minimal radiation loss as low as 0.13 dB. Moreover, a cascaded configuration of curved waveguide sections each with an alternative curvature direction was reported in Ref. [9]. This configuration is designed in such a way that polarization conversion builds up as the wave propagates along that cascaded arrangement of the curved waveguides.

Due to the radiation losses, great efforts have been directed toward the design of a single-section polarization converter that contains no transition losses and has shorter device length. This approach depends on slanting one of the waveguide sidewalls [9, 10] or the use of curved optical waveguides [11]. However, the compact PRs based on the curved optical waveguides [11] rely on a very small radius of curvature that will not be easy for fabrication. In addition, the PRs with slanted sidewall [9, 10] require a complex fabrication process including dry and wet etching techniques.

The reported polarization rotation devices in the optical frequency band are mostly composed of electro-optic material that utilizes the anisotropic property of the materials [12]. Further, a mode-evolution-based PR structure was proposed [13] by twisting a waveguide, which causes the rotation of the optical axes. However, a complicated process including E-beam lithography is required to achieve the precise dimensional control of asymmetric bi-level tapers [13]. Therefore, a deep etched width taper connected to a directional coupler was suggested by Mertens *et al.* [14] to solve this problem. A mode-evolution-based polarization rotator–splitter built on an indium phosphide (InP) substrate was reported by combining a mode converter and an adiabatic asymmetric Y-coupler [15]. A novel concept for a polarization splitter–rotator based on silicon nanowires was investigated by Dai and Bowers [16]. A polarization splitter and rotator based on a tapered directional coupler with relaxed fabrication tolerance was reported and demonstrated on the silicon-on-insulator platform by Ding *et al.* [17]. The surface plasmon polariton (SPP) approach is used to further scale down the size of an optical PR for a silicon-on-insulator (SOI) waveguide. In this regard, an ultracompact SPP-based PR of device length 3 μm and −11 dB polarization extinction ratio is reported [18]. However, the SPP PR [18] suffers from loss of 12 dB due to the SPP effect; therefore, it is difficult to improve the extinction ratio.

PCF, with a microstructure of air holes around a solid central core, has been shown to have potential for polarization conversion [19–22]. A passive rectangular core region PCF PR with a slanted sidewall [20] has been investigated with a device length of 3102 μm. Moreover, passive silica PCF PR with an L-shaped core region was studied and analyzed in Ref. [21] with nearly 100% polarization conversion ratio and device lengths of 1743 μm. A compact triangular lattice silica PCF PR with a central hole of device length 206 μm was been presented in Ref. [22]. In addition, passive PR of device length 96 μm based on soft glass equiangular spiral PCF has been reported by Hameed *et al.* [23]. The photonic bandgap silica PCF filled with LC has also been shown experimentally to have potential for polarization conversion [24, 25]. Scolari *et al.* [24] demonstrated a PR by using a silica core PCF infiltrated with a dual-frequency LC, while Wei *et al.* [25] have introduced PCF PR filled with a negative LC.

In this chapter, a highly tunable PR based on the soft glass PCF infiltrated by NLC material (NLC-PCF) [3] is introduced and analyzed. The suggested design depends on using soft glass and NLC of types SF57 (lead silica) and E7, respectively. The refractive index of the SF57 material is higher than the ordinary n_o and extraordinary n_e refractive indices of the E7 material, which guarantees the index guiding of the light through the suggested NLC-PCF PR. The presented PR with the NLC material permits high tunability with temperature or an external electric field. This is due to the change in the orientation of the NLC molecules with an external electric field or by the temperature effect. In addition, the ordinary and extraordinary refractive indices of the NLC change with temperature that affects the modal properties of the NLC-PCF PR. The effects of the structure geometrical parameters and rotation angle of the director of the NLC on the device length and polarization efficiency are investigated. The effects of the deformation of the circular holes into elliptical holes and shifting the positions of the holes are also discussed. The influences of the temperature and operating wavelength on the PR performance are also studied.

An ultracompact PR based on silica PCF with LC core [4] is also introduced and analyzed using FVFD approaches [5, 6]. The analyzed parameters of the suggested PR are the conversion length, modal hybridness, power conversion, and CT. Further, the fabrication tolerance analysis of the reported design is investigated in detail. The proposed PR has an ultracompact device length of 4.085 μm and an almost 100% polarization conversion ratio.

8.3   Practical Applications of PRs

The huge growth in the telecommunication market has led to a dramatically increasing demand for high-density, ultrafast, and efficient systems. All-optical systems are believed to be capable of meeting the high performances required by current and next-generation telecommunications. Polarization converters play an important role in the effective operation of modern optoelectronic systems. For example, in one application we can use polarization converters to design polarization-independent components by placing the polarization converters in the middle of a component so that the wave traverses half of the component in one polarization state and the second half in the other polarization state. Further, PRs can also be used to control the polarization states. They can be used in polarization controllers to compensate polarization mode dispersion, in polarization modulators to avoid polarization hole burning in transoceanic fiber links, in polarization switches for polarization division multiplexing, or in polarization diversity heterodyne receivers to provide constant 45° operation. In semiconductor optoelectronic systems, hybrid optical modes are able to convert power between the polarization states if the modes are nearly phase matched and have an enhanced overlap. However, in order to obtain this conversion, some means of discontinuity should exist along the propagation direction, such as junctions, tapers, or bends.

Photonic crystal technology has the extreme ability to control the flow of light in a fashion not possible with conventional optical technology. The photonic crystal [26], as shown in Figure 8.1, has many physics-related applications, such as sensors, prisms, medical devices, and accurate terahertz imaging that can be used in accurate diagnostics. Most of these applications need couplers, switches, PRs, and polarization splitters. The proposed PCF PR can be used as part of a comprehensive all-PCF-based system [27], where on the same PCF platform, many functionalities can be realized. The PCF-based system combines different types of waveguides,

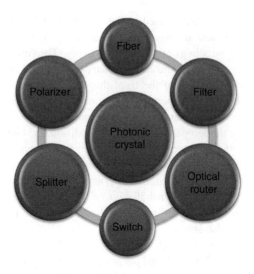

Figure 8.1   Photonic crystal-based applications

that is, channel waveguides, coupled cavity waveguides, PRs, and splitters to connect the different functions of the circuit. Being all fabricated on the same PCF platform, the integration between various PCF-based devices and the proposed PR will be straightforward. In this regard, various LC-infiltrated PCFs have been reported, showing high birefringence-tunable PCFs [28], and efficient tunable control of transmission windows [29]. In addition, liquid crystal photonic crystal fiber (LC-PCF) can be used to design all-in-fiber components, such as broadband polarimeters [30], PRs [23, 31], and tunable filters [32]. These vital elements in optical communication or sensing systems cover a wide range of applications to stabilize the operation of polarization-sensitive devices, such as modulators, interferometers, and fiber laser systems. In addition, these elements can be used to suppress polarization CT and dispersion and can add an extra degree of freedom in nonlinear applications.

8.4   Operation Principles of PRs

In the design of optical waveguides for PR structures, it is necessary that at the outset the magnitude of the nondominant field components of the fundamental transverse electric (TE) and transverse magnetic (TM) polarized modes increases and, more significantly, that their profiles be modified to increase the overlap with the dominant field components. Next, the offsets are arranged between the uniform waveguide sections such that a significant polarization rotation can take place at each junction, along with a low insertion loss. Since the propagation constants for the TE- and TM-polarized waves are different, it is also necessary to correct the phase mismatch at regular intervals. If the length of each waveguide section matches the half-beat length of the two polarized modes, constructive interference may lead to complete polarization conversion. Here the half-beat length L_π is defined as follows:

$$L_\pi = \frac{\pi}{\beta_{\text{TE}} - \beta_{\text{TM}}} \tag{8.1}$$

Here, β_{TE} and β_{TM} are the propagation constants of the quasi-TE and -TM polarized modes, respectively. It should be noted that for a single-section PR, the length of the waveguide with the hybrid modes is set to the half-beat length L_π to reverse the polarization state. It has been reported [10] that it is possible to achieve polarization rotation in a single-section design using a slanted waveguide. In this case, it is important to design a polarization rotating waveguide (PRW) that can support the quasi-TE and -TM modes, which are highly hybrid with nearly equal field components. The degree to which these modes are hybrid (the "hybridness") is such that the effective polarization angles are about 45° from the vertical or horizontal axes for the two polarization states. In such a case, if a nearly pure TE mode is introduced from a standard input waveguide (IW) that supports modes with very little degree of hybridization, this incident mode excites both the quasi-TE and -TM modes of almost similar modal amplitudes. However, as the two modes propagate along the PRW, they are out of phase and their combined modal fields produce a nearly pure field. At this position, this TM mode requires collection by an output waveguide (OW), which supports a nearly pure TM mode. It should be noted that the sidewalls of the slanted PR may not be vertical, particularly when wet etching is used, and such slanted walls can affect the polarization coupling [10]. Therefore, the suggested single-section highly tunable NLC-PCF PR will be easier to fabricate than the PR with slanted sidewalls.

Another approach to a polarization-converter design that has not yet been extensively studied relies on the use of ultrashort optical waveguide bends. It has been shown experimentally [33] that an arrangement of cascaded bend sections with opposite curvatures can play the role of a polarization converter at the expense of considerable radiation loss and the creation of a lengthy device. The same sort of cascaded bend arrangement has been investigated theoretically using the indirect coupled-mode approach. This approach is based on scalar-wave modal solution of the optical waveguide bend that uses the FDM [34].

When a pure TE mode is introduced into a bent waveguide section, it excites two hybrid modes of the waveguide bend. Hence, initially the TM power is nearly zero and as the wave propagates along the bend section, the coupling between the two hybrid TE and TM modes gives rise to a polarization rotation, which reaches its peak value at a rotation angle equivalent to half the beat length of these two modes. In simulating the cascaded configuration, it is very important to determine the arc or segment angle of each waveguide bend section in such a way that a constructive interference between the TE and TM modes is allowed to occur. In this approach, the segment angle is defined as the angle at which a maximum polarization conversion occurs in a single section. Beyond this arc angle, if the wave is allowed to propagate a further distance, and due to the improper phase matching between the TE and TM modes, the conversion process reverses. Hence, at this particular position of the bend section, if another bend with opposite curvature is connected, the process of polarization rotation builds up. Therefore, a careful study of polarization conversion in such a single waveguide bend section is essential.

8.5 Numerical Simulation Strategy

To account accurately for polarization rotation, it is essential to use full-vectorial numerical approaches. As a first step toward the simulation of a polarization converter, we need to accurately determine the modal solution of the input straight waveguide. Many modal solution techniques have been proposed in the past 20 years; however, the vector *H*-field finite-difference-based modal solution [5] has been proved to be one of the most powerful, versatile, and yet numerically efficient numerical techniques. Because the modes in general optical waveguides are hybrid in nature, all six field components exist, and the use of such a vector modal solution technique is not only justified but also mandatory. In this chapter, the FVFDM [5] with perfect matched layer boundary conditions is used to find the full-vectorial quasi-TE and -TM modes for the input and PR waveguides. Because the present modal solution is based on the use of the *H*-field, which is continuous around both the dielectric interfaces and the corners, the singularity and/or the discontinuity problems of the normal *E*-field components around the dielectric corners and the interfaces are completely avoided.

To simulate the effect of propagation in the suggested PR, a more suitable approach, such as the BPM, is necessary. Most available BPM algorithms are based on the finite differences that rely on rectangular or even uniform transverse meshes [6]. Also, many BPM approaches [6, 35–38] are used to solve either the scalar or the semi-vectorial wave equations. However, owing to the hybrid nature of the modes of the optical-waveguide bend, a FV BPM approach has to be used to account for the coupling between the TE and TM modes, which is responsible for polarization rotation. In this chapter, we use a FV BPM based on the numerically efficient FDM [6]. In this approach, the formulation is based on the TM field components in which the zero-divergence condition is automatically satisfied, so as to not permit spurious waves to propagate. Also, a rigorous perfectly matched layer boundary condition is included, in a form suitable for finite-difference applications, to mimic a nearly reflectionless boundary to the undesired radiation waves.

8.6   Design of NLC-PCF PR

Figure 8.2 shows a cross section of the suggested NLC-PCF PR [3]. The reported PR is a triangular lattice soft glass PCF whose cladding holes have been infiltrated with an NLC of type E7. The background material of the NLC-PCF is a soft glass of type SF57 (lead silica). All the holes have the same diameter d and are arranged with a hole pitch Λ.

The ordinary n_o and extraordinary n_e refractive indices of the E7 material were measured previously by Li *et al.* [39] at different visible wavelengths in the temperature range from 15

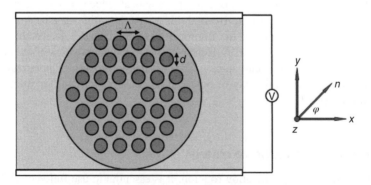

Figure 8.2   Cross section of the NLC-PCF sandwiched between two electrodes and surrounded by silicone oil. The director of the NLC is shown at the right. *Source*: Ref. [3]

to 50°C with a step of 5°C. Then, the Cauchy model was used to fit the measured n_o and n_e which are described as follows [39]:

$$n_e = A_e + \frac{B_e}{\lambda^2} + \frac{C_e}{\lambda^4}. \tag{8.2}$$

$$n_o = A_o + \frac{B_o}{\lambda^2} + \frac{C_o}{\lambda^4}. \tag{8.3}$$

Here, A_e, B_e, C_e, A_o, B_o, and C_o are the coefficients of the Cauchy model. The Cauchy coefficients at $T=25°C$ are $A_e = 1.6933$, $B_e = 0.0078\,\mu m^2$, $C_e = 0.0028\,\mu m^4$, $A_o = 1.4994$, $B_o = 0.0070\,\mu m^2$, and $C_o = 0.0004\,\mu m^4$. The variations of n_o and n_e of the E7 material with the wavelength at different temperatures T from 15 to 50°C with a step of 5°C are shown in Figure 8.2. It is evident from the figure that n_e is higher than n_o at the measured temperature values within the reported wavelength range. In the proposed design, the relative permittivity tensor ε_r of the E7 material is taken as [40] follows:

$$\varepsilon_r = \begin{pmatrix} n_o^2 \sin^2\varphi + n_e^2 \cos^2\varphi & (n_e^2 - n_o^2)\cos\varphi\sin\varphi & 0 \\ (n_e^2 - n_o^2)\cos\varphi\sin\varphi & n_o^2 \cos^2\varphi + n_e^2 \sin^2\varphi & 0 \\ 0 & 0 & n_o^2 \end{pmatrix}. \tag{8.4}$$

Here, φ is the rotation angle of the director of the NLC with respect to x-axis as shown in Figure 8.2.

The proposed in-plane alignment of the NLC can be exhibited under the influence of appropriate homeotropic anchoring conditions [28, 29, 40]. In this regard, Haakestad et al. [28], demonstrated experimentally that, in the strong field limit, the NLC of type E7 is aligned in-plane in capillaries of diameter 5 μm. In addition, Alkeskjold and Bjarklev [41] presented experimentally in-plane alignment of the E7 material in PCF capillaries of diameter 3 μm, with three different rotation angles, 0°, 45°, and 90°, and using two sets of electrodes.

The background of the NLC-PCF PR is a soft glass of type SF57 (lead silica). The wavelength-dependent refractive index of the SF57 material is also shown in Figure 8.3. It is revealed from the figure that the refractive index of the SF57 material is greater than n_o and n_e of the E7 material, which guarantees the index guiding of the light through the high index core NLC-PCF. The Sellmeier equation of the soft glass of type SF57 [42] is given by

$$n_{SF57}^2 = A_o + A_1\lambda^2 + \frac{A_2}{\lambda^2} + \frac{A_3}{\lambda^4} + \frac{A_4}{\lambda^6} + \frac{A_5}{\lambda^8}, \tag{8.5}$$

where n_{SF57} is the refractive index of the SF57 material, $A_o = 3.24748$, $A_1 = -0.0096\,\mu m^{-2}$, $A_2 = 0.0494\,\mu m^2$, $A_3 = 0.00294\,\mu m^4$, $A_4 = -1.4814\times10^{-4}\,\mu m^6$, and $A_5 = 2.7843\times10^{-5}\,\mu m^8$ [42].

In this study, the hole pitch Λ and d/Λ ratio are taken as 5.0 and 0.7 μm, respectively. In addition, ordinary and extraordinary refractive indices of the E7 material are fixed to 1.5024 and 1.6970, respectively, at the operating wavelength $\lambda=1.55$ μm and at a temperature of

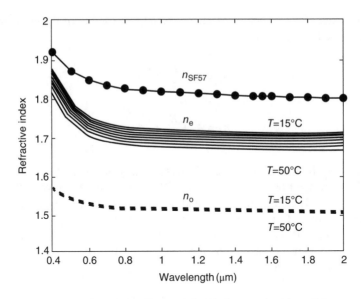

Figure 8.3 Variation of n_o and n_e of the E7 material with the wavelength at different temperatures T from 15 to 50°C with a step of 5°C. The solid line with closed circles represents the variation of the refractive index of the SF57 material n_{SF57} with the wavelength. *Source*: Ref. [3]

25°C. Moreover, the refractive index n_{SF57} of the SF57 material is taken as 1.802 at the operating wavelength 1.55 μm.

The NLC-PCF supports highly hybrid modes, as will be shown in the next section that is very useful in designing polarization conversion devices. In addition, the expected overlap between the vector field components of the quasi-TE and -TM modes can improve the polarization conversion through the NLC-PCF. However, only a small amount of mode conversion can take place in the soft glass PCF with air holes because the dominant and nondominant field profiles of the two polarized modes are not of very unequal amplitudes. In addition, the dominant field profile is symmetric in nature, while the nondominant field profile is antisymmetric. This type of PCF with very little hybridization can be used as an input or output waveguide.

8.7 Numerical Results

8.7.1 Hybridness

The degree to which the modes of the NLC-PCF are hybrid can be affected by the rotation angle of the NLC; therefore, the effect of the rotation angle on the modal hybridness and the conversion length are investigated thoroughly. Figure 8.4 shows the variation of the conversion length and hybridness for the quasi-TE and -TM modes, with the rotation angle, φ [3]. However, the hole pitch Λ and d/Λ ratio are taken as 5.0 and 0.7 μm, respectively. In this study, the hybridness is defined as follows:

$$\text{Hybridness} = \frac{\max|H_u|}{\max|H_v|} \tag{8.6}$$

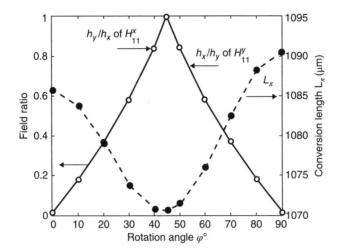

Figure 8.4 Variation of the field component ratios and the conversion length with the rotation angle of the director of the NLC of type E7. The hole pitch Λ and d/Λ ratio are taken as 5.0 and 0.7 μm, respectively. *Source*: Ref. [3]

Here, u and v are x and y for the quasi-TE mode and y and x for the quasi-TM mode. It is evident from Figure 8.4 that as φ is increased from 0° to 45°, the hybridness of the modes is increased, to reach a maximum value of 0.9978 and 0.9982 for the quasi-TE and -TM modes, respectively, at a rotation angle of 45°. Therefore, complete polarization conversion can occur at $\varphi = 45°$. However, if φ is further increased, the modal hybridness is reduced from its maximum value. Also, it is observed that the modal hybridness for the quasi-TM mode is approximately equal to that of the quasi-TE mode. Figure 8.4 reveals that the rotation angle has a slight effect on the conversion length. This is due to the slight effect of the rotation angle on the birefringence. Initially, the conversion length decreases slightly from 1086 to 1072 μm on increasing the rotation angle from 0° to 45°. Then the conversion length increases slightly to 1090 μm on increasing the rotation angle to 90°.

8.7.2 Operation of the NLC-PCF PR

As shown in the inset of Figure 8.5, when a TE-polarized mode obtained from the soft glass PCF with air holes is launched directly into the NLC-PCF, the input power excites two hybrid modes along the suggested PR waveguide. These two modes become out of phase at a distance L_π from the beginning of the PR section. Therefore, the H_y component will be canceled while the H_x component will be added, which produces a nearly pure TM mode. The calculated L_π using the FVFD-BPM is 1072 μm which is in an excellent agreement with 1070.54 μm calculated by the FVFDM. The polarization power factors, P_y and P_x, are defined as the power carried by the H_y and H_x field components, respectively, over the PR waveguide cross section, normalized to the total power. Figure 8.5 shows the variations of the P_x power for the TE input along the axial direction, at different rotation angles of the director of the NLC, while the hole pitch Λ and d/Λ ratio are fixed to 5.0 and 0.7 μm, respectively. It is evident from the figure that, for the TE excitation, initially P_x is zero, and it slowly increases to a maximum value at $z = L_\pi$

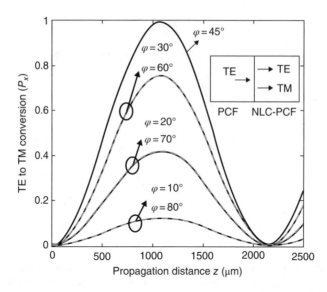

Figure 8.5 Evolution of the TM powers along the propagation direction at different rotation angles of the director of the NLC. The solid black lines represent P_x at $\varphi = 10°$, $20°$, $30°$, and $45°$, while the values of P_x at $\varphi = 60°$, $70°$, and $80°$ are represented by dotted lines. The hole pitch Λ and d/Λ ratio are taken as $5.0\,\mu m$ and 0.7, respectively. *Source:* Ref. [3]

and if the PR section is not terminated at this position the optical power P_x starts decreasing. It should be noted that nearly 99.81% polarization conversion can be obtained at $z = 1072\,\mu m$ when the rotation angle is $45°$. It is also revealed from Figure 8.5 that the conversion ratio increases with increasing the rotation angle from $0°$ to $45°$ until complete conversion occurs at $\varphi = 45°$. Then the conversion ratio decreases with increasing rotation angle, which is very compatible with the behavior of the modal hybridness, as shown in Figure 8.4. Additionally, the modal hybridness and hence the conversion ratios when $\varphi = 10°$, $20°$, and $30°$ are approximately equal to the hybridness, and hence the conversion ratios at $\varphi = 80°$, $70°$, and $60°$, respectively, as shown in Figures 8.4 and 8.5.

Figure 8.6 shows the field distributions of the field components H_y and H_x of the quasi-TE mode at different waveguide sections z, 0, 268, 536, 804, and $1072\,\mu m$. It is evident from the figure that, at $z = 0$, the quasi-TE mode of the soft glass PCF with air holes is launched into the NLC-PCF core and, as the propagation distance increases the power of the quasi-TM increases and that of the quasi-TE mode decreases. At $z = 268\,\mu m$, the powers of the quasi-TE and -TM modes normalized to the total power are 0.852 and 0.148, respectively. However, at $z = 804\,\mu m$, the normalized powers of the quasi TE and TM modes are 0.147 and 0.853, respectively. Finally, at $z = 1072\,\mu m$, the power is almost completely transferred from the quasi-TE mode to the quasi-TM mode with a negligible fraction in the quasi-TE mode.

The CT can be defined as the unwanted oppositely polarized power, normalized to the total input power, which remains at the end of the PR waveguide. For example, if a TE-polarized power is incident, then most of the power will be converted into TM-polarized power at the end of the PR section, although there might be some amount of power remaining. The remaining power, which is normalized to the total input power, is referred to as the CT. In this study,

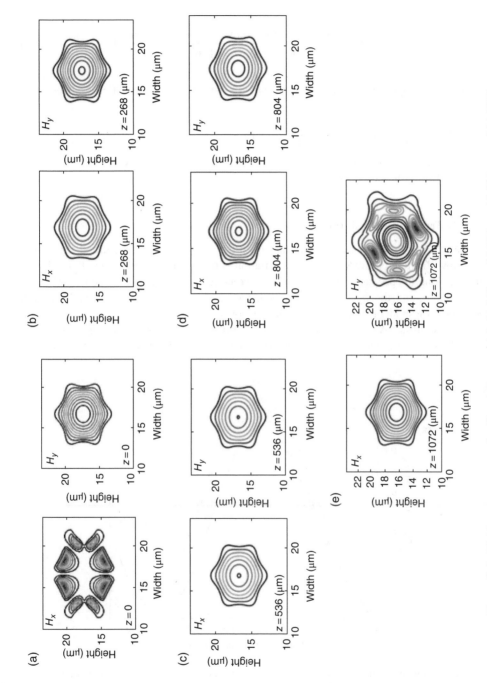

Figure 8.6 Field contour patterns for H_y and H_x at $z=$(a) 0, (b) 268 μm, (c) 536 μm, (d) 804 μm, and (e) 1072 μm

the CT is a measure of the unwanted TE power, or P_y, remaining at the end of the PR which can be expressed as follows:

$$CT = 10 \log_{10} \frac{\text{Undesired output power of the TE mode}}{\text{Desired output power of the TM mode}} \text{ dB} \qquad (8.7)$$

The CT of the suggested PR calculated by the FVFD-BPM is equal to $-26.90\,\text{dB}$ at $\varphi = 45°$.

8.7.3 Effect of Structure Geometrical Parameters

In this section, the effects of the structure geometrical parameters on the performance of the suggested PR are studied in detail.

8.7.3.1 Effect of the d/Λ Ratio

First, the effects of the d/Λ ratio at constant Λ on the conversion length and the conversion ratio are investigated. Figure 8.7 shows the variation of the TM-polarized powers P_x for the TE excitation along the propagation direction at different d/Λ ratios, 0.55, 0.6, and 0.7, while Λ, φ, temperature, and the operating wavelength are fixed to $2\,\mu\text{m}$, $45°$, $25°\text{C}$, and $1.55\,\mu\text{m}$, respectively. It is evident from the figure that the conversion length decreases with increasing d/Λ ratio at constant Λ, while the conversion ratio increases. The maximum powers of the quasi-TM mode at d/Λ ratios, 0.55, 0.6, and 0.7 are 0.965, 0.973, and 0.981, respectively.

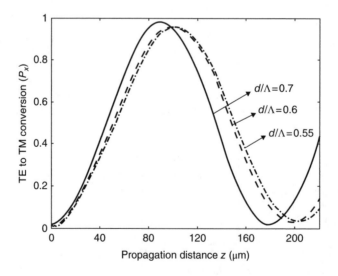

Figure 8.7 Evolution of the TM power P_x for the TE excitation along the propagation direction for different d/Λ ratios, 0.55, 0.6, and 0.7, while the Λ, φ, temperature and wavelength are fixed to $2\,\mu\text{m}$, $45°$, $25°\text{C}$, and $1.55\,\mu\text{m}$, respectively. *Source*: Ref. [3]

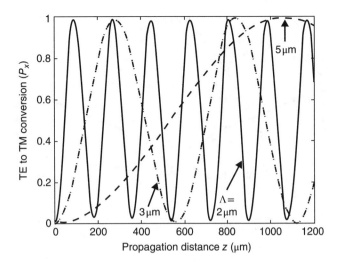

Figure 8.8 Evolution of the TM powers P_x for the TE excitation along the propagation direction at different hole pitch Λ values, while the d/Λ ratio, φ, temperature, and wavelength are fixed to 0.7, 45°, 25°C, and 1.55 μm, respectively. *Source*: Ref. [3]

8.7.3.2 Effect of the Hole Pitch Λ

The influence of the hole pitch Λ at constant d/Λ ratio is the next parameter to be studied. Figure 8.8 shows the evolution of the TM powers P_x for the TE excitation along the propagation direction at different hole pitch Λ, 2, 3, and 5 μm, while d/Λ ratio, φ and the operating wavelength are taken as 0.7, 45°, and 1.55 μm, respectively. It is revealed from the figure that the conversion length and the maximum value of P_x increase with increasing the hole pitch at constant d/Λ. The maximum values of P_x at $\Lambda = 2$, 3, and 5 μm are found to be 0.981, 0.994, and 0.998,. The corresponding conversion lengths are 91, 281, and 1072 μm, respectively. The NLC-PCF PR with $\Lambda = 5$ μm, $d/\Lambda = 0.7$, $T = 25$°C, and $\varphi = 45°$ offers a high polarization conversion ratio of 99.81% with a device length of 1072 μm. Thus, this design will be considered as the appropriate device in the subsequent simulations.

8.7.4 Tolerance of the NLC Rotation Angle

It is of great importance to consider the fabrication tolerances and their effects on the polarization converter performance. The variation of the rotation angle of the director of the NLC is the first parameter to be considered. To study the fabrication tolerances of the device length, the power conversion and the corresponding CT for the TE excitation at specified longitudinal positions have been considered for various rotation angles φ, while the hole pitch Λ and d/Λ are fixed to 5.0 and 0.7 μm, respectively. Figure 8.9 shows the variation of the TM power, P_x, and the CT values, in decibels, with the rotation angle, φ, in the range from 10° to 80° at both the exact value of L_π and the specified device length of 1072 μm. It is found that the behavior of the P_x variation with the rotation angle is very compatible with the behavior of the modal hybridness, as shown in Figure 8.4. As has been shown before, the hybridness is a maximum at $\varphi = 45°$, the value of P_x reaches its maximum value at $\varphi = 45°$, and also the corresponding

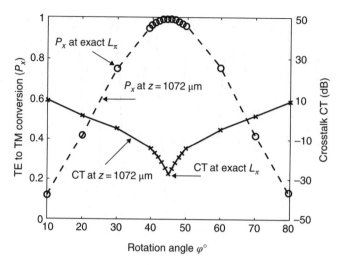

Figure 8.9 Variation of the converted power P_x for the TE excitation and the crosstalk at $z = L_\pi$ and $z = 1072\,\mu m$ with the rotation angle of the director of the NLC. *Source*: Ref. [3]

CT is a minimum value of $-27\,dB$. It is evident from the figure that changing φ yields large variation in the TM power, P_x as well as the corresponding CT. In addition, the TM power, P_x, and the CT values, in decibels, at both the exact value of L_π and at the specified device length of $1072\,\mu m$ are approximately equal. These results are compatible with the small effect of changing φ on the conversion length, as seen in Figures 8.4 and 8.5. It should be noted that when fabricating the device within the angular range of the rotation angle from $42°$ to $48°$, the maximum P_x power and the CT at the designed length will always be better than 0.987 and $-19\,dB$, respectively. However, for φ ranges from $10°$ to $40°$ and from $50°$ to $80°$, the maximum value of P_x will decrease to 0.119 and the CT will deteriorate to $8.69\,dB$. The tolerance and tunability of the rotation angle of the NLC can be achieved with a good accuracy in a strong field limit [28] with stets of electrodes as successfully experimentally reported in Refs. [28, 41].

8.7.5 Tolerance of Structure Geometrical Parameters

The tolerance of the structure geometrical parameters, such as the d/Λ ratio, holes' shape, holes' positions, and variation of the ordinary and extraordinary refractive indices of the NLC with temperature are also studied.

8.7.5.1 Tolerance of the d/Λ Ratio

First, the fabrication tolerances of the device length, the power conversion, and the corresponding CT at specified longitudinal positions were considered for various d/Λ ratios while the hole pitch Λ is constant at $5.0\,\mu m$. For this purpose, the d/Λ ratio was varied from 0.58 to 0.74, while the optimum device length was fixed at $1072\,\mu m$, as it corresponds to the ideal in design conditions, $\Lambda = 5.0\,\mu m$, $d/\Lambda = 0.7$, $T = 25°C$, and $\varphi = 45°$. As shown in Figure 8.10, the value of P_x for the TE excitation at the corresponding values of L_π slightly increases from

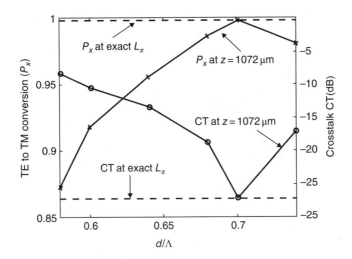

Figure 8.10   Variation of the converted power P_x for the TE excitation and the crosstalk at $z = L_\pi$ and $z = 1072\,\mu m$ with the d/Λ ratio, while Λ, φ, and temperature are fixed to $5\,\mu m$, $45°$, and $25°C$, respectively. *Source*: Ref. [3]

0.9980 to the maximum value of 0.9981 as d/Λ increases from 0.58 to 0.74. On the other hand, if the power conversion is to be calculated at a fixed device length of $1072\,\mu m$, the P_x values will be much lower than those at the exact values of L_π, except at $d/\Lambda = 0.7$, where the exact value of L_π is itself $1072\,\mu m$. The CT measured at the specified device length of $1072\,\mu m$ decreases from -8.36 to $-27.28\,dB$ as d/Λ is increased from 0.58 to 0.7. Then the CT increases to $-17.03\,dB$ as d/Λ is increased to 0.74. It is also important to mention that for the d/Λ range from 0.69 to 0.73, the CT value will still be less than $-20\,dB$. However, if the d/Λ ratio changes to 0.58, the maximum value of P_x will decrease to 0.873 and the CT will deteriorate to $-8.36\,dB$.

8.7.5.2   Tolerance of the Hole Shape

Next, the effect of the deformation of the circular holes into elliptical holes on the PR performance is investigated. Here, the ellipticity is defined as the ratio (a/b) where, a and b are the radii in the x- and y-directions, respectively, of the elliptical holes, as shown in the inset of Figure 8.11. Figure 8.11 shows the variation of the hybridness and the conversion length with the ellipticity ratio. As the ellipticity ratio increases from 0.8 to 1.0, the hybridness of the quasi-TE modes increases from 0.974 to 0.998. Then the hybridness decreases to 0.988 when the ellipticity ratio increases to 1.2. In addition, the conversion length L_π shows a reduction from 1223 to $946\,\mu m$ over the range of ellipticity ratio from 0.8 to 1.2.

Figure 8.12 shows the variation of the converted TM power P_x for the TE excitation and the CT for both at exact L_π and at the designed device length $1072\,\mu m$ with the ellipticity a/b ratio. It is revealed from the figure that a maximum power conversion occurs when $a/b = 1$ because the hybridness is a maximum at this ratio. As, a/b is varied away from 1.0, P_x at $1072\,\mu m$ is always lower than P_x at L_π due to L_π mismatch. It can also be seen from the figure that fabricating

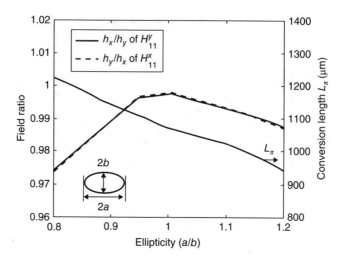

Figure 8.11   Variation of the field component ratios and the conversion length with the ellipticity ratio a/b

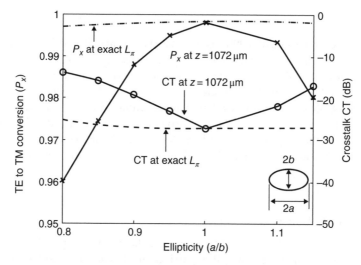

Figure 8.12   Variation of the converted power P_x for the TE excitation and the crosstalk at $z=L_\pi$ and $z=1072\,\mu m$ with the ellipticity ratio a/b. *Source*: Ref. [3]

the device for a tolerance of ellipticity ratio equal to 1.0 ± 0.1, the CT at the designed length will always be better than $-19\,dB$. The effect of the b/a ratio on the PR performance is also considered. It is found that the b/a has the same effect as a/b on the PR performance.

8.7.5.3   Tolerance of the Hole Position

It is well known that the holes of the first ring have a great impact on the PCF characteristics; therefore, the effect of shifting the position of one of the holes of the first ring on the PR

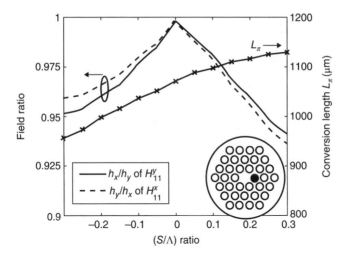

Figure 8.13 Variation of the conversion length and the field components with the shift ratio in the x-direction. *Source*: Ref. [3]

performance will be considered next. Initially, it is found that if the centers of all holes of the first ring are shifted in the x- or y-directions, this will have a slight effect on the PR performance because the core region will be unchanged. Therefore, the effect of shifting only one hole is studied. The solid black hole in the first ring of the NLC-PCF shown in the inset of Figure 8.13 is the one under consideration in this study. Figure 8.13 shows the variation of the conversion length with the shift ratio (S/Λ) where S is the shift distance in the positive or negative x-direction. The variation of the hybridness for the two polarized modes with the shift ratio in the x-direction is also shown in Figure 8.13. It should be noted that the positive and negative shift ratio indicates the shift in the positive and negative x-direction, respectively. In this study, the other parameters are kept constant at $\Lambda=5.0\,\mu m$, $d/\Lambda=0.7$, $T=25°C$, $\varphi=45°$, and $\lambda=1.55\,\mu m$. It is noticed that the conversion length increases with increasing the shift ratio in the positive x-direction while, the conversion length decreases with increasing the shift ratio in the negative x-direction. However, the hybridness decreases with increasing the shift ratio in the positive or negative x-direction. The conversion length changes from 1072 to 1129 μm and 958 μm for the shift ratio of 0.3 in the positive and negative x-directions, respectively. In addition, the hybridness difference over the shift ratio range from 0 to 0.3 is considerably small, giving a value of 0.06 and 0.05 for the shift in the positive and negative x-directions, respectively.

In order to investigate the tolerance in the position of one of the holes in the first ring over the polarization rotation, the converted TM power P_x for the TE excitation, at $z=L_\pi$ and at $z=1072\,\mu m$ and the CT are obtained within the range of shift ratio from 0 to 0.3 in the positive and negative x-directions, as shown in Figure 8.14. It is revealed from Figure 8.14 that as the shift ratio increases from 0 to 0.3, the converted power P_x at exact L_π reduces from 0.998 to 0.994 for the shift in the positive x-direction and from 0.998 to 0.996 for the shift in the negative x-direction. This is due to the decrease in the hybridness with increasing shift ratio, as shown in Figure 8.13. These 0.004 and 0.002 differences in the P_x slightly affect the behavior of the PR. In addition, the converted power P_x at $z=1072\,\mu m$ also does not show much

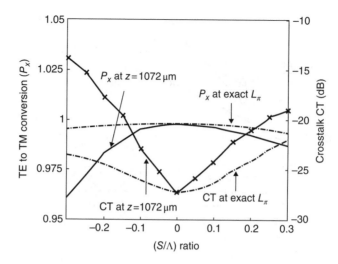

Figure 8.14 Variation of the converted power P_x for the TE excitation and the crosstalk at $z=L_\pi$ and $z=1072\,\mu m$ with the shift ratio in the x-direction. *Source*: Ref. [3]

deviation with the shift in the positive x-direction from the P_x curve at exact L_π, because L_π changes only by 57 μm along the whole shift range considered. However, for the shift in the negative x-direction, L_π changes by 114 μm; therefore, P_x at $z=1072\,\mu m$ shows more deviation from the P_x curve at exact L_π as shown in Figure 8.14. The effect of the shift in the positive y-direction on the performance of the PR has been studied, and it is found to have the same effect as the shift in the positive x-direction. It is worth noting that by fabricating the device for a tolerance of a shift ratio of 0.15 in the positive x- or y-directions, the crosstalk at the designed length will always be better than −19 dB.

8.7.6 Tolerance of the Temperature

The ordinary and extraordinary refractive indices of the NLC of type E7 change with temperature variation, which affects the performance of the reported PR. Therefore, the effect of the variation of the ordinary and extraordinary refractive indices of the NLC with temperature on the hybridness and the device length will be the next parameter to be considered. In this study, the other parameters are still invariant at $\Lambda=5.0\,\mu m$, $d/\Lambda=0.7$, $\lambda=1.55\,\mu m$, and $\varphi=45°$. Figure 8.15 shows the variation of the hybridness for both modes and the conversion length with the temperature. It is evident from the figure that the hybridness of the two polarized modes decreases slightly with increasing temperature, while the conversion length increases significantly. The conversion length increases from 935 to 1596 μm as the temperature increases from 15 to 45°C. In order to investigate the tolerance in the temperature difference over the polarization rotation, the converted TM power P_x for the TE excitation, at $z=L_\pi$ and at $z=1072\,\mu m$ and the CT are obtained within the T range from 15 to 50°C, as shown in Figure 8.16. It is observed from this figure that the CT at exact L_π decreases from −26 to −32 dB on increasing the temperature from 15 to 50°C. However, the CT at $z=1072\,\mu m$ decreases from −12.44 to −27.277 dB as T increases from 15 to 25°C then the CT increases to −1.33 dB

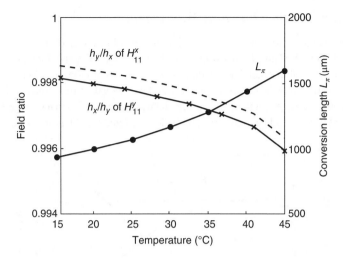

Figure 8.15 Variation of the field component ratios for the two polarized modes and the conversion length with temperature

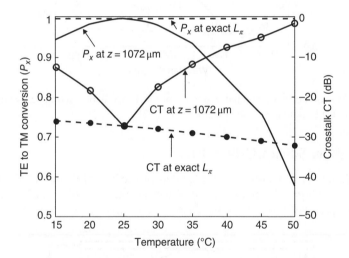

Figure 8.16 Variation of the converted power P_x for the TE excitation and the crosstalk at $z=L_\pi$ and $z=1072\,\mu m$ with temperature. *Source*: Ref. [3]

as the temperature increases to 50°C. In addition, there is a large difference between P_x at $z=L_\pi$ and at $z=1072\,\mu m$ due to the significant effect of the temperature on the conversion length. It can also be seen from Figure 8.16 that by fabricating the device for a tolerance of 25 ± 4°C, the CT at the designed length will always be better than -19 dB. The temperature tolerance can be fulfilled by using a thermo-electric module as described experimentally by Woliński *et al.* [43, 44] allowing temperature control in the 10–120°C range with 0.1°C long-term stability and electric field regulation in the 0–1000 V range with frequencies from 50 Hz to 2 kHz.

8.7.7 Tolerance of the Operating Wavelength

The effect of the operating wavelength, λ, variation on the performance of the PR is also studied in the range from 1.53 to 1.63 μm. In this study, the change in the refractive index of the composing materials with wavelength is taken into account. Figure 8.17 shows the variation of the TM power, P_x for the TE excitation, and the corresponding CT at the exact values of L_π and the design length of 1072 μm. At the given device length of 1072 μm, if the wavelength varies in the range 1.55 ± 0.02 μm, the maximum value of P_x will decrease from 0.998 to 0.997 and the CT is about −25 dB.

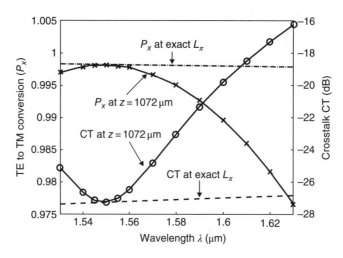

Figure 8.17 Variation of the converted power P_x for the TE excitation and the crosstalk at $z = L_\pi$ and $z = 1072$ μm with the operating wavelength. *Source*: Ref. [3]

8.8 Ultrashort Silica LC-PCF PR

Figure 8.18 shows a cross section of the suggested ultracompact LC silica PCF (LC-SPCF) [4] PR. The reported design is based on silica PCF with an NLC core of type E7. The cladding air holes of the proposed PCF with the same diameter d are arranged in a hexagonal shape with a

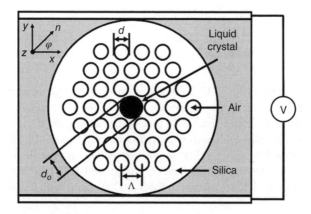

Figure 8.18 Cross section of the suggested LC-SPCF PR. *Source*: Ref. [4]

hole pitch Λ. A central hole of diameter d_o filled with NLC is inserted in the core region where $d_o > d$. The n_o and n_e of the NLC are given by Ref. [39]. In this study, the modal hybridness is expressed by the overlap integral [45] between the vector fields for the TE mode in the input section $E(\text{TE},1)$ and for the TM mode in the proposed LC-SPCF PR section as follows:

$$S = \iint \left(E_x \left(\text{TE},1 \right) E_x \left(\text{TM},2 \right) + E_y \left(\text{TE},1 \right) E_y \left(\text{TM},2 \right) \right) dxdy. \tag{8.8}$$

The effect of the rotation angle φ on the performance of the suggested PR is first introduced at the operating wavelength $\lambda = 1.55\,\mu\text{m}$ and $T = 25°\text{C}$. In this study, the infiltrated central hole radius $= 1.65\,\mu\text{m}$, $d/\Lambda = 0.6$, and $\Lambda = 2.5\,\mu\text{m}$. The variation of the φ-dependent modal hybridness is shown in Figure 8.19. The inset figures show the field profiles of the fundamental TM mode at the central line of the proposed PR waveguide at $\varphi = 45°$ and $90°$. It is revealed from Figure 8.19 that the modal hybridness and φ are directly proportional when $\varphi = 0$–$45°$. However, they are inversely proportional when $\varphi = 45$–$90°$. Therefore, the modal hybridness has minimum values of 2.183×10^{-6} and 2.017×10^{-6} at $\varphi = 0°$ and $90°$, respectively. Consequently, the suggested structure can be used as a PR device at $\varphi = 45°$ where maximum modal hybridness is achieved.

In this study, the conventional silica PCF without a central hole is used as input and output waveguides to launch the TE-polarized mode shown in Figure 8.20a and b and to collect the required polarization at the outset. The splice between the conventional PCF and the proposed LC-SPCF can be achieved successfully by the techniques reported by Wang *et al.* [46] and Hu *et al.* [47] without destroying the NLC state. The TE mode is launched into the suggested LC-SPCF PR with $\varphi = 45°$. Figure 8.20 shows contour plots of the H_x and H_y field components calculated by the FVFD-BPM at different locations in the axial directions $z = 0$, $L_\pi/2$, and L_π. As may be seen from the figure, the launched TE mode is converted to TM mode at $z = L_\pi$.

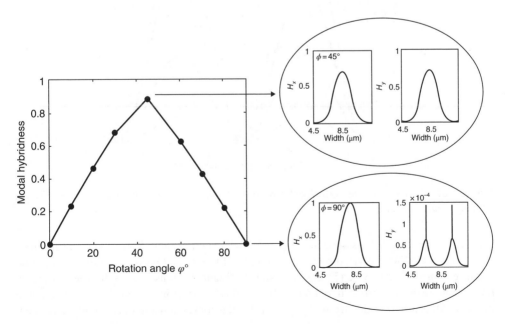

Figure 8.19   Variation of the φ-dependent modal hybridness. *Source*: Ref. [4]

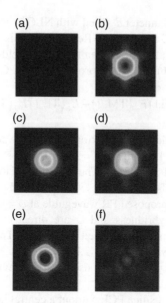

Figure 8.20 Contour plot of the H_x and H_y field components at $z=0$, $L_\pi/2$, and L_π. (a) H_x at $z=0$; (b) H_y at $z=0$; (c) H_x at $z=L_\pi/2$; (d) H_y at $z=L_\pi/2$; (e) H_x at $z=L_\pi$; and (f) H_y at $z=L_\pi$. *Source*: Ref. [4]

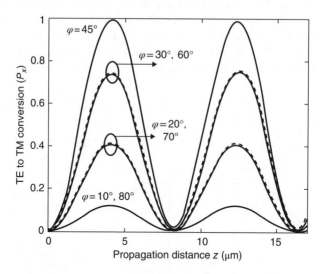

Figure 8.21 Variation of the normalized TM power P_x with the propagation distance at different φ values. *Source*: Ref. [4]

The conversion length L_π is 4085 nm at $\varphi =45°$ as shown in Figure 8.21. It shows variations of the converted TM polarized power P_x normalized to the total power with the propagation distance, at $\varphi= 10°, 20°, 30°, 45°, 60°, 70°$, and $80°$. Figure 8.21 also shows that the conversion ratio and φ are directly proportional for $\varphi=0–45°$. If φ is further increased, the conversion ratio decreases. It is also revealed from Figure 8.21 that the conversion length is slightly dependent

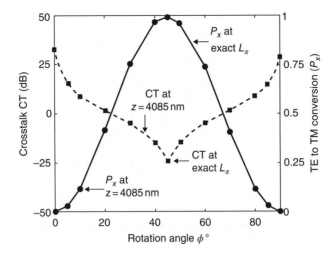

Figure 8.22 The effect of the rotation angle on the crosstalk and converted power P_x at $z=L_\pi$ and $z=4085$ nm. *Source*: Ref. [4]

on the rotation angle. Moreover, the modal hybridness and hence the conversion ratios when $\varphi=10°$, $20°$, and $30°$ are approximately equal to those at $\varphi=80°$, $70°$, and $60°$, respectively, as shown in Figures 8.19 and 8.21.

Figure 8.22 shows the rotation angle, φ-dependent TM power, P_x, and the CT values, in decibels at the designed device length of 4085 nm and at the exact value of L_π. It is revealed from the figure that the value of P_x is a maximum at $\varphi=45°$ and the corresponding CT is a minimum value of -23.975 dB. In addition, the conversion length is slightly dependent on the rotation angle, as shown in Figure 8.21; therefore, the TM powers and the CTs at $z=L_\pi$ and at 4085 nm are nearly the same. It also shows that the TM power, P_x and the corresponding CT are sensitive to the rotation angle variation. The numerical results reveal that the suggested PR has a tolerance of $\varphi=45° \pm 3°$ at which the maximum P_x power and the CT at the designed device length are better than 0.9848 and -18 dB, respectively. The control and tunability of the rotation angle of the NLC can be experimentally achieved with high accuracy using sets of electrodes and a strong field limit [28].

The behavior of the proposed PR is temperature-dependent due to the temperature-dependent ordinary n_o and extraordinary n_e refractive indices of the NLC [39]. In this investigation, the NLC central hole radius is fixed to 1.65 μm and the other parameters are still invariant at $\Lambda=2.5$ μm, $d/\Lambda=0.6$, and $\lambda=1.55$ μm. Figure 8.23a shows the effect of the temperature on the modal hybridness and conversion length. As may be seen, the modal hybridness and conversion length are directly proportional to the temperature. Figure 8.23b shows the effect of temperature variation on the converted TM power P_x and the CT at $z=L_\pi$ and $z=4085$ nm. As the temperature increases from 15 to 50°C, the CT at exact L_π decreases from -23.766 to -24.955 dB, as shown in Figure 8.23b. However, the CT at $z=4085$ nm is a minimum at $T=25°$C. The conversion length increases from 3853.64 to 4825.71 nm on increasing the temperature from 15 to 50°C. Therefore, the converted power P_x and CT at $z=L_\pi$ are different form those at the designed device length $z=4085$ μm. It is also revealed from Figure 8.23b that the CT at $z=4085$ nm is better than -20 dB for $T=25\pm5°$C. In this

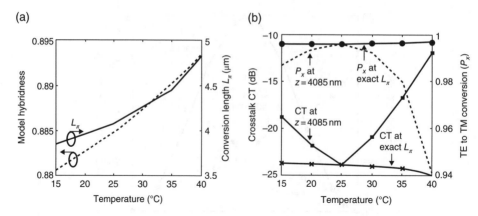

Figure 8.23 The effect of the temperature on (a) the modal hybridness and conversion length and (b) the crosstalk and converted power P_x at $z=L_\pi$ and $z=4085$ nm. *Source*: Ref. [4]

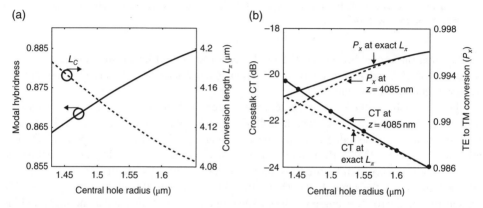

Figure 8.24 The effect of the NLC central hole radius on (a) the modal hybridness and conversion length and (b) the crosstalk and converted power P_x at $z=L_\pi$ and $z=4085$ nm. *Source*: Ref. [4]

regard, Woliński *et al.* [44] used a thermo-electric module to control the temperature in the 10–120°C range with 0.1°C long-term stability.

The influence of the NLC central hole radius is next investigated while the other parameters are taken as $\Lambda=2.5\,\mu$m, $d/\Lambda=0.6$, $T=25$°C, $\varphi=45°$, and $\lambda=1.55\,\mu$m. Figure 8.24a shows the effect of the NLC central hole radius on the modal hybridness and conversion length. The numerical results reveal that the NLC effect and hence the modal hybridness increase by increasing the central hole radius. As the central hole radius increases, the difference between the propagation constants of the quasi-TE and quasi-TM modes increases, and hence the conversion length slightly decreases as shown in Figure 8.24a. The conversion length changes slightly from 4186 to 4085 nm within the central hole range from 1.43 to 1.65 µm. Therefore, the difference between the converted power and CT at $z=L_\pi$ and at $z=4085$ nm is small as shown in Figure 8.24b. Figure 8.24b shows the effect of the central hole radius on the converted TM power P_x and the CT at $z=L_\pi$ and $z=4085$ nm. It is also evident from this figure that the CT at $z=4085$ nm is better than −20 dB within the central hole radius range from 1.43 to 1.65 µm.

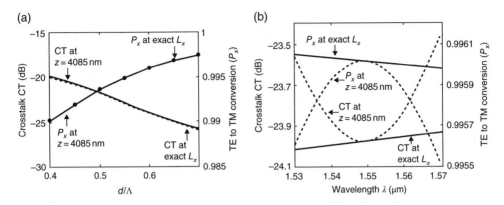

Figure 8.25 The effect of (a) the d/Λ ratio and (b) the wavelength on the crosstalk and P_x at $z=L_\pi$ and $z=4085$ nm. *Source*: Ref. [4]

The effect of the d/Λ ratio at constant hole pitch Λ of 2.5 µm is also reported while the other parameters are fixed to central hole radius $=1.65$ µm, $T=25°C$, $\varphi=45°$, and $\lambda=1.55$ µm. It is found that the conversion length decreases slightly from 4142 to 4053 nm by increasing the d/Λ ratio from 0.4 to 0.7, respectively. Therefore, the difference between the CT and P_x at $z=L_\pi$ and at $z=4085$ nm is very small, as shown in Figure 8.25a. Figure 8.25 shows the effect of (a) the d/Λ ratio and (b) the wavelength on the P_x and the CT at $z=L_\pi$ and $z=4085$ nm. Throughout the d/Λ ratio range from 0.4 to 0.7, the maximum P_x power and the CT at $z=4085$ nm is better than 0.99015 and -20 dB, respectively.

The effect of the operating wavelength variation is finally tested taking into account the wavelength-dependent refractive indices of the composing materials. As shown from Figure 8.25b, the operating wavelength has a tolerance of $\lambda=1.55\pm0.02$ µm at which the CT is better than -23.47 dB.

8.9 Fabrication Aspects of the NLC-PCF PR

The silica NLC-PCF PR [4] depends on the triangular lattice silica PCF widely fabricated by the stack-and-draw method [48]. Due to the dramatic improvement in the PCF fabrication process, more complex PCF structures can now be fabricated. In this regard, PCF with a hole pitch of 2.3 µm or less can be fabricated successfully with a 0.3 µm of hole diameter [48]. In addition, more complex PCF can be realized, such as V-shaped high birefringence PCFs [49]. It should also be noted that the PCFs with hole pitch of 2.0 µm and hole diameter 1.0 µm or less (0.7–1.0 µm) infiltrated with NLC have been fabricated successfully by Woliński *et al.* [50]. In addition, the infiltration of only central holes has been suggested in Ref. [51] using a similar filling technique with similar dimensions to our proposed PCF. A two-photon direct-laser writing technique now offers full flexibility in selective filling of PCFs [52]. Huang *et al.* [53] have also reported a selective filling technique that depends on filling the large central hole faster than the smaller cladding holes. Hence, the PCF can be cleaved with the required length at that position at which only the central hole is filled with the infiltrated material. Therefore, it is believed that the suggested design can be experimentally fulfilled. Through the filling process, the LC can be filled at the isotropic temperature [54] which decreases the filling time. After 3 h, Scolari [54] has reported that the filling length is sufficient to be used in communication system devices, such as PR and splitters. Therefore, the authors believe that the filling time of our proposed PCF will not be different than that reported by Scolari [54], a relatively long PCF can be followed by a cleaving process to create the required shorter length as reported by Huang *et al.* [53].

8.10   Summary

In this chapter, an overview of the previously reported PRs is given, followed by the practical applications of PRs. The operation principles of PRs and a numerical simulation strategy for studying PRs are presented. Highly tunable single-section polarization converters based on the soft glass NLC-PCF and silica LC-PCF [4] are presented and results reported. The simulation results are obtained using the FVFDM [5] and confirmed by the FVFD-BPM [6]. In addition, the influences of the different structure geometrical parameters, rotation angle of the director of the NLC, temperature and operating wavelength on the PR' performance are investigated. Finally, the fabrication aspects of the proposed PRs are presented.

References

[1] Heismann, F. and Smith, R.W. (1994) High-speed polarization scrambler with adjustable phase chirp. *IEEE Select. Top. Quantum Electron.*, **2**, 311–318.

[2] Morita, I., Tanka, K., Edagawa, N., and Suzuki, M. (1999) 40 Gb/s single-channel soliton transmission over transoceanic distances by reducing Gordon-Haus timing jitter and soliton-soliton interaction. *J. Lightwave Technol.*, **2**, 2506–2511.

[3] Hameed, M.F.O. and Obayya, S.S.A. (2010) Analysis of polarization rotator based on nematic liquid crystal photonic crystal fiber. *Lightwave Technol. J.*, **28**, 806–815.

[4] Hameed, M.F.O. and Obayya, S.S.A. (2014) Ultrashort silica liquid crystal photonic crystal fiber polarization rotator. *Opt. Lett.*, **39**, 1077–1080.

[5] Fallahkhair, A.B., Li, K.S., and Murphy, T.E. (2008) Vector finite difference modesolver for anisotropic dielectric waveguides. *J. Lightwave Technol.*, **26** (11), 1423–1431.

[6] Huang, W.P. and Xu, C.L. (1993) Simulation of three-dimensional optical waveguides by a full-vector beam propagation method. *IEEE J. Quantum Electron.*, **29** (10), 2639–2649.

[7] Shani, Y., Alferness, R., Koch, T. *et al.* (1991) Polarization rotation in asymmetric periodic loaded rib waveguides. *Appl. Phys. Lett.*, **59**, 1278–1280.

[8] Obayya, S.S.A., Rahman, B.M.A., Grattan, K.T.V., and El-Mikati, H.A. (2001) Beam propagation modelling of polarization rotation in deeply etched semiconductor bent waveguides. *IEEE Photon. Technol. Lett.*, **13**, 681–683.

[9] Obayya, S.S.A., Somasiri, N., Rahman, B.M.A., and Grattan, K.T.V. (2003) Full vectorial finite element modeling of novel polarization rotators. *Opt. Quantum Electron.*, **35**, 297–312.

[10] Rahman, B.M.A., Obayya, S.S.A., Somasiri, N. *et al.* (2001) Design and characterisation of compact single-section passive polarization rotator. *IEEE J. Lightwave Technol.*, **19**, 512–519.

[11] Obayya, S.S.A., Rahman, B.M.A., Grattan, K.T.V., and El-Mikati, H.A. (2001) Improved design of a polarization converter based on semiconductor optical waveguide bends. *Appl. Opt.*, **40**, 5395–5401.

[12] Reinhart, F.K., Logan, R.A., and Sinclair, W. (1982) Electrooptic polarization modulation in multielectrode AlxGa1-xAs rib waveguides. *Quantum Electron. IEEE J.*, **18**, 763–766.

[13] Watts, M.R. and Haus, H.A. (2005) Integrated mode-evolution-based polarization rotators. *Opt. Lett.*, **30**, 138–140.

[14] Mertens, K., Opitz, B., Hovel, R. *et al.* (1998) First realized polarization converter based on hybrid supermodes. *Photon. Technol. Lett. IEEE*, **10**, 388–390.

[15] Yuan, W., Kojima, K., Wang, B. *et al.* (2012) Mode-evolution-based polarization rotator-splitter design via simple fabrication process. *Opt. Express*, **20**, 10163–10169.

[16] Dai, D. and Bowers, J.E. (2011) Novel concept for ultracompact polarization splitter-rotator based on silicon nanowires. *Opt. Express*, **19**, 10940–10949.

[17] Ding, Y., Liu, L., Peucheret, C., and Ou, H. (2012) Fabrication tolerant polarization splitter and rotator based on a tapered directional coupler. *Opt. Express*, **20**, 20021–20027.

[18] Jing, Z., Shiyang, Z., Shiyi, C. *et al.* (2011) An ultracompact surface plasmon polariton-effect-based polarization rotator. *Photon. Technol. Lett. IEEE*, **23**, 1606–1608.

[19] Hameed, M.F.O. and Obayya, S.S.A. (2011) Polarization rotator based on soft glass photonic crystal fiber with liquid crystal core. *Lightwave Technol. J.*, **29**, 2725–2731.

[20] Hameed, M.F.O. and Obayya, S.S.A. (2011) Design of passive polarization rotator based on silica photonic crystal fiber. *Opt. Lett.*, **36**, 3133–3135.

[21] Hameed, M.F.O., Obayya, S.S.A., and El-Mikati, H.A. (2012) Passive polarization converters based on photonic crystal fiber with l-shaped core region. *Lightwave Technol. J.*, **30**, 283–289.

[22] Hameed, M.F.O., Abdelrazzak, M., and Obayya, S.S.A. (2013) Novel design of ultra-compact triangular lattice silica photonic crystal polarization converter. *Lightwave Technol. J.*, **31**, 81–86.

[23] Hameed, M.F.O., Heikal, A.M., and Obayya, S.S.A. (2013) Novel passive polarization rotator based on spiral photonic crystal fiber. *Photon. Technol. Lett. IEEE*, **25**, 1578–1581.

[24] Scolari, L., Alkeskjold, T., Riishede, J. *et al.* (2005) Continuously tunable devices based on electrical control of dual-frequency liquid crystal filled photonic bandgap fibers. *Opt. Express*, **13**, 7483–7496.

[25] Wei, L., Eskildsen, L., Weirich, J. *et al.* (2009) Continuously tunable all-in-fiber devices based on thermal and electrical control of negative dielectric anisotropy liquid crystal photonic bandgap fibers. *Appl. Opt.*, **48**, 497–503.

[26] Inoue, K. and Ohtaka, K. (2003) *Photonic Crystals*, Springer, New York.

[27] Shinoj, V.K. and Murukeshan, V.M. (2012) Hollow-core photonic crystal fiber based multifunctional optical system for trapping, position sensing, and detection of fluorescent particles. *Opt. Lett.*, **37**, 1607–1609.

[28] Haakestad, M.W., Alkeskjold, T.T., Nielsen, M. *et al.* (2005) Electrically tunable photonic bandgap guidance in a liquid-crystal-filled photonic crystal fiber. *IEEE Photon. Technol. Lett.*, **17** (4), 819–821.

[29] Zografopoulos, D.C., Kriezis, E.E., and Tsiboukis, T.D. (2006) Photonic crystal-liquid crystal fibers for single-polarization or high-birefringence guidance. *Opt. Express*, **14**, 914–925.

[30] Wei, L., Alkeskjold, T.T., and Bjarklev, A. (2010) Tunable and rotatable polarization controller using photonic crystal fiber filled with liquid crystal. *Appl. Phys. Lett.*, **96** (24), 241104–241104-3.

[31] Hameed, M.F.O., Obayya, S.S.A., and Wiltshire, R.J. (2010) Beam propagation analysis of polarization rotation in soft glass nematic liquid crystal photonic crystal fibers. *Photon. Technol. Lett. IEEE*, **22**, 188–190.

[32] Du, J., Liu, Y., Wang, Z. *et al.* (2008) Electrically tunable Sagnac filter based on a photonic bandgap fiber with liquid crystal infused. *Opt. Lett.*, **33**, 2215–2217.

[33] van Dam, C., Spiekman, L.H., van Ham, F.P.G.M. *et al.* (1996) Novel compact polarization converters based on ultra short bends. *IEEE Photon. Technol. Lett.*, **8**, 1346–1348.

[34] Lui, W.W., Hirono, T., Yokoyama, K., and Huang, W.P. (1998) Polarization rotation in semiconductor bending waveguides: a coupled-mode theory formulation. *Lightwave Technol. J.*, **16**, 929–936.

[35] Obayya, S.S.A., Rahman, B.M.A., and El-Mikati, H.A. (2000) New full vectorial numerically efficient propagation algorithm based on the finite element method. *J. Lightwave Technol.*, **18** (3), 409–415.

[36] Montanari, E., Selleri, S., Vincetti, L., and Zoboli, M. (1998) Finite element full vectorial propagation analysis for three dimensional z-varying optical waveguides. *J. Lightwave Technol.*, **16**, 703–714.

[37] Chund, Y., Dagli, N., and Thylen, L. (1991) Explicit finite difference vectorial beam propagation method. *Electron. Lett.*, **27** (23), 2119–2121.

[38] Huang, W.P., Xu, C.L., and Chaudhuri, S.K. (1992) A finite-difference vector beam propagation method for three-dimensional waveguide structures. *IEEE Photon. Technol. Lett.*, **4**, 148–151.

[39] Li, J., Wu, S.T., and Brugioni, S. (2005) Infrared refractive indices of liquid crystals. *J. Appl. Phys.*, **97** (7), 073501–073501-5.

[40] Ren, G., Shum, P., Yu, X. *et al.* (2008) Polarization dependent guiding in liquid crystal filled photonic crystal fibers. *Opt. Commun.*, **281**, 1598–1606.

[41] Alkeskjold, T.T. and Bjarklev, A. (2007) Electrically controlled broadband liquid crystal photonic bandgap fiber polarimeter. *Opt. Lett.*, **32** (12), 1707–1709.

[42] Leong, J.Y.Y. (2007) Fabrication and applications of lead-silicate glass holey fiber for 1-1.5microns: nonlinearity and dispersion trade offs. PhD thesis. Faculty of Engineering, Science and Mathematics Optoelectronics Research Centre, University of Southampton.

[43] Woliński, T.R., Szaniawska, K., Ertman, S. *et al.* (2006) Influence of temperature and electrical fields on propagation properties of photonic liquid crystal fibers. *Meas. Sci. Technol.*, **17** (50), 985–991.

[44] Woliński, T.R., Ertman, S., Czapla, A. *et al.* (2007) Polarization effects in photonic liquid crystal fibers. *Meas. Sci. Technol.*, **18**, 3061–3069.

[45] Weinert, C.M. and Heidrich, H. (1993) Vectorial simulation of passive TE/TM mode converter devices on In P. *IEEE Photon. Technol. Lett.*, **5**, 324–326.

[46] Wang, F., Yuan, W., Hansen, O., and Bang, O. (2011) Selective filling of photonic crystal fibers using focused ion beam milled microchannels. *Opt. Express*, **19**, 17585–17590.

[47] Hu, D.J.J., Jun Long, L., Ying, C. *et al.* (2012) Fabrication and characterization of a highly temperature sensitive device based on nematic liquid crystal-filled photonic crystal fiber. *Photon. J. IEEE*, **4**, 1248–1255.

[48] Knight, J.C., Birks, T.A., Russell, P.S.J., and Atkin, D.M. (1996) All-silica single-mode optical fiber with photonic crystal cladding. *Opt. Lett.*, **21**, 1547–1549.

[49] Wojcik, J., Mergo, P., Makara, M. *et al.* (2010) V type high birefringent PCF fiber for hydrostatic pressure sensing. *Photon. Lett. Poland*, **2**, 10–12.

[50] Woliński, T.R., Szaniawska, K., Bondarczuk, K. *et al.* (2005) Propagation properties of photonic crystal fibers filled with nematic liquid crystals. *Opto-Electron. Rev.*, **13**, 177–182.

[51] Woliński, T.R., Czapla, A., Ertman, S. *et al.* (2007) Tunable highly birefringent solid-core photonic liquid crystal fibers. *Opt. Quantum Electron.*, **39**, 1021–1032.

[52] Vieweg, M., Gissibl, T., Pricking, S. *et al.* (2010) Ultrafast nonlinear optofluidics in selectively liquid-filled photonic crystal fibers. *Opt. Express*, **18**, 25232–25240.

[53] Huang, Y., Xu, Y., and Yariv, A. (2004) Fabrication of functional microstructured optical fibers through a selective-filling technique. *Appl. Phys. Lett.*, **85** (22), 5182–5184.

[54] Scolari, L. (2009) Liquid crystals in photonic crystal fibers: fabrication, characterization and devices. PhD thesis. Department of Photonics Engineering, Technical University of Denmark.

9

Applications of Nematic Liquid Crystal-Photonic Crystal Fiber Coupler

9.1 Introduction

Due to their different uses in communication systems, the directional coupler [1–4] and its application to polarization splitters [5] and multiplexer–demultiplexer (MUX–DEMUX) [6] have attracted the interest of many researchers in recent years. In this chapter, novel designs of a polarization splitter and MUX–DEMUX based on the soft glass nematic liquid crystal-photonic crystal fiber (NLC-PCF) coupler presented in Chapter 6 are introduced and analyzed. The simulation results are calculated by the full-vectorial finite-difference method (FVFDM) [7] and FVFD beam propagation method (FVFD-BPM) [8]. In addition, the fabrication aspects of the proposed designs are described.

9.2 Multiplexer–Demultiplexer

9.2.1 Analysis of NLC-PCF MUX–DEMUX

The fiber coupler can separate two wavelengths at λ_1 and λ_2 if the coupling length $L_{ck,\lambda1}$ at the wavelength λ_1 and the coupling length $L_{ck,\lambda2}$ at the wavelength λ_2 satisfy the following coupling ratio [9]:

$$\gamma_k = L_{ck,\lambda1} : L_{ck,\lambda2} = i : j. \tag{9.1}$$

Here, i and j are two integers of different parities and k denotes the x- or y-polarization state. In this case, the length of the coupler is equal to $L_f = L_{ck,\lambda2} \times i/j$. Therefore, to achieve the shortest coupler, the optimal value of γ should be 2. Figure 9.1 shows the coupling length ratio for the transverse electric (TE) polarization of the NLC-PCF coupler shown in the inset as a

Computational Liquid Crystal Photonics: Fundamentals, Modelling and Applications, First Edition.
Salah Obayya, Mohamed Farhat O. Hameed and Nihal F.F. Areed.
© 2016 John Wiley & Sons, Ltd. Published 2016 by John Wiley & Sons, Ltd.

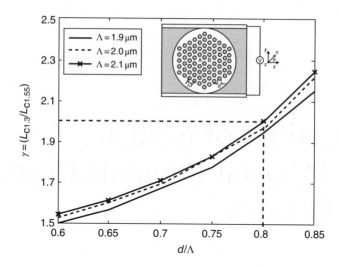

Figure 9.1 Variation of the coupling length ratio for the quasi-TE modes of the NLC-PCF with the d/Λ ratio at different hole pitch values $\Lambda = 1.9, 2.0$, and $2.1\,\mu m$. *Source*: Ref. [6]

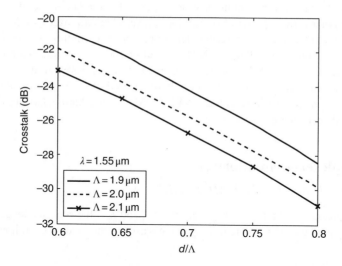

Figure 9.2 Variation of the crosstalk for the quasi-TE modes of the NLC-PCF at $\lambda = 1.55\,\mu m$ with the d/Λ ratio at different hole pitch values Λ, 1.9, 2.0, and $2.1\,\mu m$. *Source*: Ref. [6]

function of the d/Λ ratio at different hole pitch values, 1.9, 2.0, and $2.1\,\mu m$ with $\lambda_1 = 1.3\,\mu m$ and $\lambda_2 = 1.55\,\mu m$. In this study, the rotation angle of the director of the NLC and the temperature are taken as 90° and 25°C, respectively. It is found that the coupling length ratio γ increases, and hence the crosstalk (CT) decreases with increasing d/Λ ratio at a given hole pitch, as seen from Figures 9.1 and 9.2. The CT is a measure of the unwanted power, or the power of unwanted wavelength, remaining at the end of the NLC-PCF coupler. Figure 9.2 shows the variation of the CT for the quasi-TE mode at $\lambda = 1.55\,\mu m$ with the d/Λ ratio at different hole pitch values, 1.9, 2.0, and 2.1 μm. The CT of $\lambda = 1.3$ μm is not shown in Figure 9.2 for clarity and the numerical results reveal that the CT at $\lambda = 1.3$ μm is lower than that at $\lambda = 1.55$ μm due to the modes being well confined through the core regions at shorter wavelength. As can be seen from Figure 9.1, the coupling length ratio equals 2.0019 at $d/\Lambda = 0.8$ and a hole pitch of 2.1 μm. The coupling lengths calculated by the FVFDM are equal to 3264.63 and 1630.71 μm at $\lambda = 1.3$ and 1.55 μm, respectively.

9.2.2   Beam Propagation Study of the NLC-PCF MUX–DEMUX

The FVFD-BPM is applied to confirm the validity of the suggested design [6]. Initially, at $z = 0$, the fundamental components of the quasi-TE mode H_y of soft glass PCF with air holes obtained using the FVFDM [7] at $\lambda = 1.3$ and 1.55 μm are launched into the left core A of the NLC-PCF coupler. These input fields, in turn, start to transfer to the right core B of the coupler and at the corresponding coupling lengths the fields are almost completely transferred to the right core. The coupling lengths calculated by the FVFD-BPM are equal to 3266 and 1632 μm at $\lambda = 1.3$ and 1.55 μm, respectively, which is in excellent agreements with those calculated by the FVFDM. The ratio between the coupling lengths at $\lambda = 1.3$ and 1.55 μm is slightly larger than 2.0. Therefore, the length of the proposed coupler is $L_f = [3266 + (2 \times 1632)]/2.0 = 3265$ μm at which the two wavelengths are well separated. Figure 9.3 shows the power transfers for the quasi-TE modes normalized to the input powers at wavelengths of 1.3 and 1.55 μm in the left core of the NLC-PCF coupler. As shown in the figure, the two wavelengths are separated after a propagation distance $L_f = 3265$ μm. Therefore, when launched from one end of the MUX–DEMUX, the signals at the wavelength of 1.3 μm will exit at the other core, while the signals at the wavelength of 1.55 μm can be obtained at the same core.

The field distributions of the dominant field component H_y of the quasi-TE mode at $\lambda = 1.3$ and 1.55 μm are shown in Figure 9.4 at different waveguide sections z, 0, 1632 and 3265 μm.

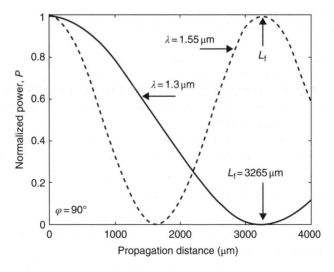

Figure 9.3   Evolution of the normalized powers at the left core A for the quasi-TE modes at two different wavelengths, 1.30 and 1.55 μm along the propagation direction. *Source*: Ref. [6]

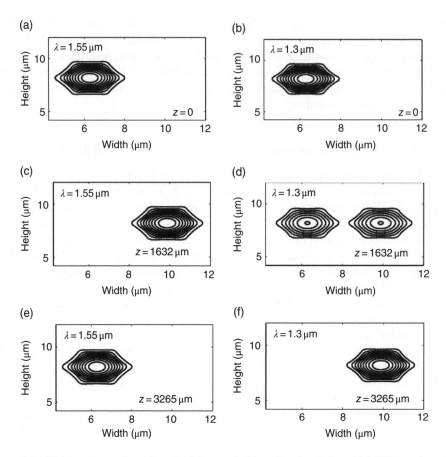

Figure 9.4 Field contour patterns for H_y of the quasi-TE modes at $z=0$ (a and b), $1632\,\mu m$ (c and d), and $3265\,\mu m$ (e and f) at $\lambda=1.3$ and $1.55\,\mu m$. *Source*: Ref. [6]

It is evident from the figure that, at $z=0$, the input fields are launched into the left core and as the propagation distance increases the power in the right core increases and that in the left core decreases. At $z=1632\,\mu m$, which is equal to the coupling length of the wavelength of $1.55\,\mu m$, the power of the quasi-TE mode at $\lambda=1.55\,\mu m$ is almost completely transferred to the right core. However, the powers of the quasi-TE mode at $\lambda=1.3\,\mu m$ in the left and right cores of the coupler normalized to the input power are equal to 0.50013 and 0.49963, respectively, at $z=1632\,\mu m$. Finally, the two wavelengths are separated after a propagation distance $L_f=3265\,\mu m$ [6].

9.2.3 CT of the NLC-PCF MUX–DEMUX

The wavelength-dependent CT between the two cores of the NLC-PCF MUX–DEMUX for the desired wavelength of $\lambda=1.3$ and $1.55\,\mu m$ are shown in Figure 9.5a and b, respectively, and are given in decibels as follows:

$$\text{CT}\left(\text{around } 1.55\,\mu m\right) = 10\log_{10}\frac{\text{Undesired normalized power of the } 1.55\,\mu m \text{ signal}}{\text{Desired normalized power of the } 1.3\,\mu m \text{ signal}}. \quad (9.2)$$

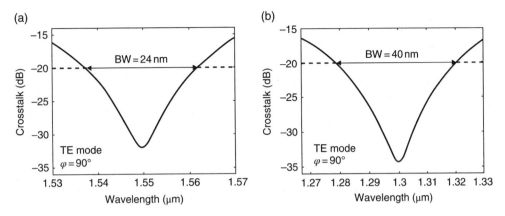

Figure 9.5 Wavelength dependence crosstalk between the two cores of the NLC-PCF coupler for the desired wavelength of (a) $\lambda = 1.3\,\mu m$ and (b) $\lambda = 1.55\,\mu m$. *Source*: Ref. [6]

$$\text{CT}\left(\text{aroud } 1.3\ \mu m\right) = 10\log_{10}\frac{\text{Undesired normalized power of the 1.3 }\mu m \text{ signal}}{\text{Desired normalized power of the 1.55 }\mu m \text{ signal}}. \quad (9.3)$$

Figure 9.5a and b shows that the proposed MUX–DEMUX has bandwidths of 24 and 40 nm around $\lambda = 1.55$ and $1.3\,\mu m$, respectively, for the CTs that are lower than −20 dB. In addition, the bandwidths are 13 and 22 nm around $\lambda = 1.55$ and $1.3\,\mu m$, respectively, for the CTs that are lower than −25 dB. Therefore, the reported MUX–DEMUX is less sensitive to the perturbation introduced during the fabrication process due to the low-level CTs with wide wavelength ranges. At $\lambda = 1.3\,\mu m$, the modes are better confined through the core regions than the modes at $\lambda = 1.55\,\mu m$. Therefore, the undesired normalized power of the signal of wavelength $1.3\,\mu m$ is less than the undesired normalized power of the signal of wavelength $1.55\,\mu m$. Consequently, the bandwidth around $\lambda = 1.3\,\mu m$ is greater than the bandwidth around $\lambda = 1.55\,\mu m$, as shown in Figure 9.5.

The bandwidths of the NLC-PCF MUX–DEMUX [6] are much larger than those reported in Refs. [3, 10]. The bandwidths in Ref. [3] are 5.1 and 2.7 nm around $\lambda = 1.3$ and $1.55\,\mu m$, respectively, while the bandwidths in [10] are 5.5 and 2.0 nm around $\lambda = 1.3$ and $1.55\,\mu m$, respectively. Further, the proposed MUX–DEMUX is shorter than those reported in Refs. [3] and [10] of lengths 15.4 and 9.08 mm, respectively. Moreover, the NLC-PCF MUX–DEMUX has a wide wavelength range comparable to the reported MUX–DEMUX in Ref. [9] which are 42.2 and 25.4 nm around $\lambda = 1.3$ and $1.55\,\mu m$, respectively. However, the MUX–DEMUX in Ref. [9] has longer length of 10.69 mm than the NLC-PCF MUX–DEMUX length.

9.2.4   Feasibility of the NLC-PCF MUX–DEMUX

The tolerances of the fiber length and rotation angle of the director of the NLC are also investigated. In this study, the tolerance of a specific parameter is calculated, while the other parameters of the proposed design are kept constant. It is found that the fiber length allows a tolerance of ±3% at which the CTs around $\lambda = 1.3$ and $1.55\,\mu m$ are still better than −20 dB. Additionally, the rotation angle of the director of the NLC has a tolerance of ±5° at which the CTs are still better than −22 dB. If the rotation angle is changed within the range 90° ± 10°, the CTs will be still better than −15 dB.

The effect of the variation of the ordinary and extraordinary refractive indices of the NLC with the temperature on the performance of the NLC-PCF MUX–DEMUX is also considered. However, the other parameters are fixed to their designed values. It is found that the CTs of the reported MUX–DEMUX around $\lambda = 1.3$ and 1.55 μm are better than −20 dB over a temperature range from 15 to 35°C. At $\varphi = 90°$, the relative permittivity tensor ε_r of the E7 material has the diagonal form $\left[n_o^2, n_e^2, n_o^2 \right]$. Therefore, ε_{yy} is more dependent on the temperature variation, while ε_{xx} is nearly invariant. As the temperature increases, ε_{yy} decreases while ε_{xx} is nearly constant. As a result, the index contrast seen by the quasi-TE mode, and hence its coupling length are almost constant with the temperature variation. Therefore, the CTs for the quasi-TE modes are approximately temperature insensitive.

9.3 Polarization Splitter

9.3.1 Analysis of the NLC-PCF Polarization Splitter

The fiber coupler can also be used to separate the two polarized states, quasi-TE and -transverse magnetic (TM) modes, at a given wavelength if the coupling lengths L_{cTE} and L_{cTM} of the quasi-TE and -TM modes, respectively, satisfy the following coupling ratio [9]:

$$\gamma = L_{cTE} : L_{cTM} = i : j. \tag{9.4}$$

Here, i, and j are two integers of different parities. The length of the coupler should be $L_f = L_{cTM} \times i/j$.

Figure 9.6 shows the coupling length ratio between the coupling lengths of the quasi-TE and -TM modes of the NLC-PCF coupler [5] as a function of the d/Λ ratio at two different hole pitch values, 3.7 and 3.9 μm at $\lambda = 1.55$ μm. In this study, the rotation angle of the director of the NLC and the temperature are taken as 90° and 25°C, respectively. It is found that the coupling length ratio γ increases, and hence the CT decreases with increasing d/Λ ratio at a given hole pitch as revealed from Figures 9.6 and 9.7. Figure 9.7 shows the variation of the CT for the quasi-TE and -TM modes at $\lambda = 1.55$ μm with the d/Λ ratio at two different hole pitch values: 3.7 and 3.9 μm. As seen from Figure 9.7, the CT of quasi-TE is lower than that of the quasi-TM mode due to the better confinement of the quasi-TE modes through the core regions than the quasi-TM mode at $\varphi = 90°$. In addition, Figure 9.6 revealed that the coupling length ratio is 2.013 at $d/\Lambda = 0.7$ and hole pitch 3.9 μm. The coupling lengths calculated by the FVFDM are 8252.084 and 4098.65 μm for the quasi-TE and -TM modes, respectively, at 1.55 μm.

9.3.2 Beam Propagation Study of the NLC-PCF Polarization Splitter

In order to confirm the polarization splitter based on the NLC-PCF coupler [5], the FVFD-BPM is used to study the propagation along its axial direction. Initially, at $z = 0$, the fundamental components H_y and H_x of the quasi-TE and -TM modes, respectively, of a soft glass PCF with air holes obtained using the FVFDM [7] at $\lambda = 1.55$ μm are launched into the left core A of the

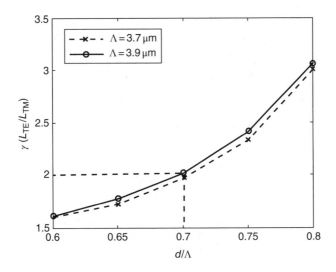

Figure 9.6 Variation of the coupling length ratio for the quasi-TE and -TM modes of the NLC-PCF coupler with the d/Λ ratio at two different hole pitch values $\Lambda = 3.7$ and $3.9\,\mu m$. *Source*: Ref. [5]

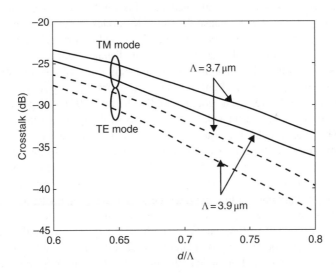

Figure 9.7 Variation of the crosstalk for the TE and TM modes of the NLC-PCF at $\lambda = 1.55\,\mu m$ with the d/Λ ratio at two different hole pitch values $\Lambda = 3.7$ and $3.9\,\mu m$. *Source*: Ref. [5]

NLC-PCF coupler. These input fields, in turn, start to transfer to the right core of the coupler and at the corresponding coupling lengths, the fields are almost completely transferred to the right core. The coupling lengths calculated by the FVFD-BPM are 8253 and $4100\,\mu m$ for the quasi-TE and -TM modes, respectively, which are in excellent agreement with those obtained by the FVFDM. The ratio of the coupling lengths L_{cTE} and L_{cTM} is slightly larger than 2.0. Therefore, the length of the proposed coupler is taken as $L_f = (8253 + 2 \times 4100)/2.0 = 8227\,\mu m$

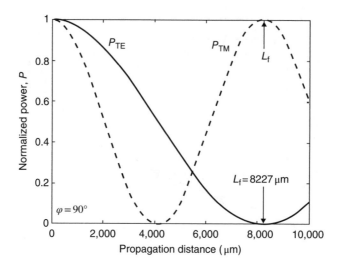

Figure 9.8 Evolution of the normalized powers at the left core A for the quasi-TE and -TM modes at the operating wavelength of 1.55 μm along the propagation direction. *Source*: Ref. [5]

at which the two polarized states are separated well. Figure 9.8 shows the power transfer normalized to the input power for the quasi-TE and -TM modes at the operating wavelength of 1.55 μm in the left core of the NLC-PCF coupler. It is evident from Figure 9.8 that the two polarized modes are separated well after a propagation distance equals $L_f = 8227$ μm. The normalized powers of the quasi-TE mode in the left and right cores of the coupler are 0.000435 and 0.9995, respectively, at $z = 8227$ μm. However, the normalized powers of the quasi-TM mode in the left and right cores of the coupler are 0.9987 and 0.0013, respectively. Therefore, when launched from one end of the splitter, the signals of the quasi-TE mode will exit at the other core B, while the signals of the quasi-TM mode will exit at the same core A.

The field distributions of the dominant field component H_y and H_x of the quasi-TE and -TM modes, respectively, at $\lambda = 1.55$ μm are shown in Figure 9.9 at different waveguide sections along the propagation direction z, 0, 4100 and 8227 μm. It is evident from this figure that, at $z = 0$, the input fields are launched into the left core and, as the propagation distance increases, the normalized power in the right core increases and that in the left core decreases. At $z = 4100$ μm, which is equal to the coupling length of the quasi TM mode, the normalized power of the quasi-pTM mode is approximately completely transferred to the right core. The normalized powers of the quasi-TM mode in the left and right cores of the coupler are 0.00096 and 0.99904, respectively. However, the normalized powers of the quasi-TE mode in the left and right cores of the coupler are 0.5051 and 0.4949, respectively, at $z = 4100$ μm. Finally, the two polarized modes are separated after a propagation distance $L_f = 8227$ μm [5].

9.3.3   CT of the NLC-PCF Splitter

The CT of the TM polarization in decibels at the left core A is given by the following equation:

$$CT_{TM} = 10 \log_{10} \frac{\text{Undesired normalized power of the TE polarization in core A}}{\text{Desired normalized power of the TM polarization in core A}}. \quad (9.5)$$

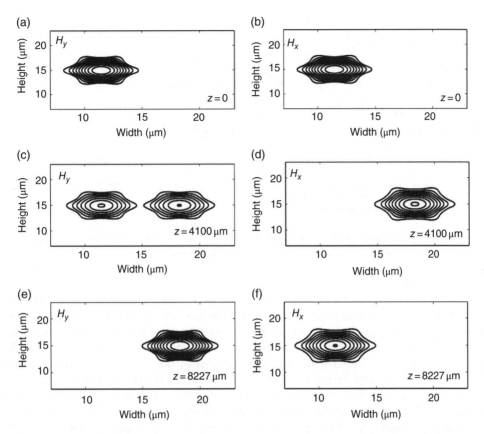

Figure 9.9 Field contour patterns for H_y and H_x of the quasi-TE and -TM modes, at $z=0$ (a and b), 4100 μm (c and d), and 8227 μm (e and f) at the operating wavelength $\lambda=1.55$ μm. *Source*: Ref. [5]

However, the CT of the TE polarization in decibels at the right core B is defined as follows:

$$\mathrm{CT_{TE}} = 10\log_{10}\frac{\text{Undesired normalized power of the TM polarization in core B}}{\text{Desired normalized power of the TE polarization in core B}}. \quad (9.6)$$

The CTs around the operating wavelength of 1.55 μm for the quasi-TE and -TM modes are shown in Figure 9.10a and b, respectively. It is revealed from this figure that the proposed splitter has bandwidths of 30 and 75 nm for the quasi-TE and -TM modes, respectively, at which the CTs are better than −20 dB. In addition, the bandwidths are 15 and 42 nm around $\lambda=1.55$ μm for the quasi-TE and -TM modes, respectively, for the CTs that are lower than −25 dB. Therefore, the reported polarization splitter is less sensitive to the perturbation introduced during the fabrication process due to the low-level CTs with wide wavelength ranges.

The bandwidths of the NLC-PCF splitter [5] are much larger than those reported in Refs. [3] and [10]. The bandwidth of the quasi-TE mode in [3] is 2.7 nm around $\lambda=1.55$ μm, while that in Ref. [10] is 2.0 nm. In addition, the proposed splitter is shorter than those reported in Refs. [3] and [10] of lengths 15.4 and 9.08 mm, respectively. Moreover, the NLC-PCF splitter has a wide wavelength range, larger than that reported for the splitter by Chen *et al.* [9] of

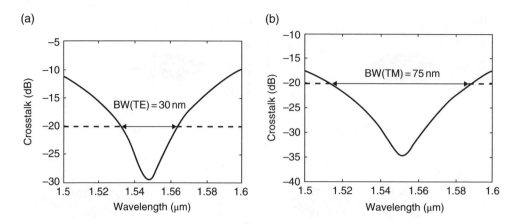

Figure 9.10 Wavelength-dependent crosstalk between the two cores of the NLC-PCF coupler around $\lambda = 1.55\,\mu m$ for (a) the quasi-TE and (b) the quasi-TM modes. *Source*: Ref. [5]

bandwidth 25.4 nm around $\lambda = 1.55\,\mu m$ for the quasi-TE mode. Furthermore, the splitter in Ref. [9] has longer length of 10.69 mm than that of the NLC-PCF splitter. It is also shown from Figure 9.10 that the bandwidth of the quasi-TE mode is less than that of the quasi-TM mode. At $\varphi = 90°$, the quasi-TE mode is more confined through the core region than the quasi-TM mode. Therefore, the quasi-TM mode is more affected by the wavelength variation around $\lambda = 1.55\,\mu m$ than the quasi-TE mode. As a result, the undesired normalized power of the quasi-TM mode at core B at the device length of 8227 μm increases with the wavelength variation around $\lambda = 1.55\,\mu m$ more than the undesired normalized power of the quasi-TE mode at core A. Therefore, the bandwidth of the quasi-TE mode at core B is less than that of the quasi-TM mode at core A.

9.3.4   Feasibility of the NLC-PCF Polarization Splitter

The tolerances of the fiber length and rotation angle of the director of the NLC are also investigated. It is worth noting that the tolerance of a specific parameter is calculated, while the other parameters of the proposed design are kept constant. It is found that the fiber length and rotation angle allow a tolerance of ±3% and ±5°, respectively, for the two polarized modes at which the CTs are still better than −20 dB.

The effect of the variation of the ordinary and extraordinary refractive indices of the NLC with temperature on the performance of the NLC-PCF splitter is also investigated. However, the other parameters of the reported design are not modified. It is found that the CT of the quasi-TM mode is better than −20 dB over a temperature range from 15 to 50°C. However, the CT of the quasi-TE mode has a tolerance of ±5°C at which the CT is better than −14 dB. This can be explained as follows. At $\varphi = 90°$, the relative permittivity tensor ε_r of the E7 material has the diagonal form $\left[n_o^2, n_e^2, n_o^2 \right]$. Therefore, ε_{yy} is more dependent on the temperature variation, while ε_{xx} is nearly invariant. As the temperature increases, ε_{yy} decreases while ε_{xx} is nearly constant. Consequently, the index contrast seen by the quasi-TM mode increases by increasing the temperature, which increases the confinement of the quasi-TM mode through the core regions. Thus, the coupling length of the quasi-TM mode increases with increasing temperature. However, the index contrast seen by the quasi-TE mode, and hence its coupling length are nearly constant with the temperature variation. Therefore, the undesired normalized power of the quasi-TM mode at core B, at the device length of 8227 µm, increases, which increases the CT of the quasi-TE mode. On the other hand, the undesired power of the quasi-TE modes at core A is approximately invariant, which has little effect on the CT of the quasi-TM mode.

9.4 Summary

Novel designs of a highly tunable MUX–DEMUX and a polarization splitter based on an index-guiding soft glass NLC-PCF coupler are introduced and analyzed by the FVFDM [7] and FVFD-BPM [8]. The numerical results reveal that the suggested polarization splitter and MUX–DEMUX can provide larger bandwidths with shorter lengths than those of the conventional silica PCF couplers with air holes. In Chapter 10, the coupling characteristics of a PCF coupler with small NLC core are discussed.

References

[1] Mangan, B.J., Knight, J.C., Birks, T.A. *et al.* (2000) Experimental study of dual core photonic crystal fiber. *Electron. Lett.*, **36**, 1358–1359.

[2] Saitoh, K., Sato, Y., and Koshiba, M. (2003) Coupling characteristics of dual-core photonic crystal fiber couplers. *Opt. Express*, **11**, 3188–3195.

[3] Florous, N., Saitoh, K., and Koshiba, M. (2005) A novel approach for designing photonic crystal fiber splitters with polarization-independent propagation characteristics. *Opt. Express*, **13** (19), 7365–7373.

[4] Zhang, L. and Yang, C. (2004) Polarization-dependent coupling in twin-core photonic crystal fibers. *J. Lightw. Technol.*, **22** (5), 1367–1373.

[5] Hameed, M.F.O. and Obayya, S.S.A. (2009) Polarization splitter based on soft glass nematic liquid crystal photonic crystal fiber. *Photon. J. IEEE*, **1**, 265–276.

[6] Sazio, P.J.A., Amezcua-Correa, A., Finlayson, C.E. *et al.* (2009) Multiplexer–Demultiplexer based on nematic liquid crystal photonic crystal fiber coupler. *Opt. Quantum Electron.*, **41**, 315–326.

[7] Fallahkhair, A.B., Li, K.S., and Murphy, T.E. (2008) Vector finite difference modesolver for anisotropic dielectric waveguides. *J. Lightw. Technol.*, **26** (11), 1423–1431.

[8] Huang, W.P. and Xu, C.L. (1993) Simulation of three-dimensional optical waveguides by a full-vector beam propagation method. *IEEE J. Quantum Electron.*, **29** (10), 2639–2649.

[9] Chen, M.Y. and Zhou, J. (2006) Polarization-independent splitter based on all-solid silica-based photonic-crystal fibers. *J. Lightw. Technol.*, **24**, 5082–5086.

[10] Florous, N.J., Saitoh, J.K., and Koshiba, M. (2006) Synthesis of polarization-independent splitters based on highly birefringent dual-core photonic crystal fiber platforms. *Photon. Technol. Lett.*, **18**, 1231–1233.

10

Coupling Characteristics of a Photonic Crystal Fiber Coupler with Liquid Crystal Cores

10.1 Introduction

Due to their different uses in communication systems, fiber couplers have attracted the interest of many researchers in recent years. They can be used to transfer, divide, or combine the optical power in communication systems. It has been shown by Mangan *et al.* [1] that it is possible to use the photonic crystal fiber (PCF) as an optical fiber coupler. In this regard, the design of PCF couplers as a polarization splitter [2, 3], a broadband directional coupler [4], wavelength division multiplex components [5, 6], and filters [7] has been proposed. Saitoh *et al.* [5] evaluated the coupling characteristics of two different dual-core PCF couplers, showing that it is possible to realize significantly shorter multiplexer–demultiplexer PCFs compared to a conventional optical fiber coupler. Florous *et al.* [2] investigated numerically the operation of a polarization-independent splitter based on the PCF with elliptical air holes. However, small bandwidths (BWs) of 5.1 and 2.7 nm are achieved around wavelengths of 1.3 and 1.55 μm, respectively, at a relatively low-level crosstalk (CT) of −20 dB. A polarization-independent splitter based on all-solid silica PCF has been proposed by Chen *et al.* [3], with a device length of 10.69 mm and BWs of 25.4 and 42.2 nm around wavelengths of 1.55 and 1.31 μm, respectively. A polarization splitter of length 20 mm based on a square-lattice PCF has been presented [8], showing CT as low as −23 dB with BWs as large as 90 nm. Recently, the authors have reported the coupling characteristics of a dual-core soft glass PCF coupler infiltrated with a nematic liquid crystal (NLC-PCF) [9]. A polarization splitter [10] and a multiplexer–demultiplexer [11] based on the soft glass NLC-PCF coupler have been introduced. The NLC-PCF splitter [10] of coupling length 8.227 mm achieved low CT better than −20 dB with great BWs of 30 and 75 nm for the quasi-transverse electric (TE) and quasi-transverse magnetic (TM), modes, respectively. The NLC-PCF multiplexer–demultiplexer [11] of length 3.265 mm also provided large BWs of 40 and 24 nm around the wavelengths of 1.3 and 1.55 μm, respectively.

Computational Liquid Crystal Photonics: Fundamentals, Modelling and Applications, First Edition.
Salah Obayya, Mohamed Farhat O. Hameed and Nihal F.F. Areed.
© 2016 John Wiley & Sons, Ltd. Published 2016 by John Wiley & Sons, Ltd.

In this chapter, a novel design of a highly tunable coupler based on soft glass PCF with air holes and dual NLC core (SGLC-PCF) [12] is presented and analyzed. The suggested design depends on using soft glass and NLC of types SF57 (lead silica) and E7, respectively. The refractive index of the SF57 material is greater than the ordinary n_o and extraordinary n_e refractive indices of the E7 material. In addition, propagation through the SGLC-PCF coupler takes place by the modified total internal reflection mechanism due to the index contrast between the core and cladding region. The SGLC-PCF coupler also has strong polarization dependence due to the infiltration of the NLC, and high tunability with temperature or an external electric field. The high tunability feature of the suggested coupler renders this design suitable for applications ranging from optical communications to biosensing and bioimaging. Arc fusion techniques [13] have been successfully implemented for the infiltration of central defect cores; therefore, the suggested design is easier for fabrication than the NLC-PCF [9] with all the cladding holes filled. The effects of the structure geometrical parameters, and temperature on the coupling characteristics of the reported coupler, are investigated. The numerical results reveal that the SGLC-PCF coupler [12] can be used as a polarization splitter of length 6232 μm, with low CT better than −20 dB, with great BWs of 250 and 60 nm around the operating wavelength of 1.55 μm for the quasi-TE and quasi-TM modes, respectively. The analysis is carried out using the full-vectorial finite-difference method (FVFDM) [14] together with the FVFD beam propagation method (FVFD-BPM) [15].

Through all simulations, the transverse step sizes are fixed to $\Delta x = \Delta y = 0.05$ μm, while the longitudinal step size Δz is taken as 1 μm; the reference index n_o, which is used to satisfy the slowly varying envelope approximation of the FVFD-BPM [15], is taken as the effective index of the fundamental mode launched at the input waveguide; and α is chosen within the range, $0.5 \le \alpha \le 1$, at which the FVFD-BPM is unconditionally stable.

10.2   Design of the PCF Coupler with LC Cores

Figure 10.1 shows a cross section of the suggested triangular lattice dual-core SGLC-PCF coupler [12]. The two identical cores of diameter d_o have been infiltrated with an NLC of type E7. All the cladding air holes have the same diameter d and are arranged with a hole pitch Λ. The separation between the two identical cores is equal to $\sqrt{3}\Lambda$. The background material of the reported SGLC-PCF coupler is a soft glass of type SF57 (lead silica).

The fiber is placed between two electrodes allowing the arbitrary control of the alignment of the NLC director via an external voltage, as shown schematically in Figure 10.1. Two silica rods with appropriate diameter are used to control the spacing between the electrodes and the fiber is surrounded by silicone oil, which has a higher dielectric strength than air [16]. Therefore, the external electric field will be uniform across the fiber cross section, which results in good alignment of the director of the NLC with constant rotation angle φ. The non-uniform electric field region will only be found at the edges, far away from the core region where the light will be propagating. As a result, the proposed coupler overall performance will not be affected by the nonuniform field distribution at the edges. Other layouts, such as those described in Refs. [17, 18] can also be used to ensure better field distribution uniformity over the fiber cross section. Wei *et al.* [19] proved that by using sets of electrodes and controlling them independently, the direction of the electrical field is rotatable under an effective driving voltage of 50 V_{rms}.

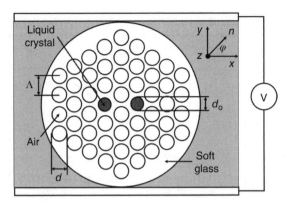

Figure 10.1   Cross section of the dual-core SGLC-PCF coupler sandwiched between two electrodes and surrounded by silicone oil. *Source*: Ref. [12]

10.3   Numerical Results

In the suggested design, the cladding air holes have the same diameter d and are arranged with a hole pitch $\Lambda = 2.0\,\mu m$, while the radius of the infiltrated NLC cores r_o is taken as $0.5\,\mu m$; the rotation angle of the director of the NLC and the temperature are fixed to 90 and 25°C, respectively; n_o, n_e, n_{SF57} are taken as 1.5024, 1.6970, and 1.802, respectively, at the operating wavelength $\lambda = 1.55\,\mu m$. The refractive index of the SF57 material is greater than n_o and n_e of the E7 material, which guarantees the index guiding of the light through the high index core SGLC-PCF coupler [12]. In this study, the power is launched at the left core and the coupling length can be obtained using the operating wavelength λ, and effective indices of the even n_{eff_e} and odd modes n_{eff_o} as follows:

$$L_C = \frac{\lambda}{2\left(n_{eff\_e} - n_{eff\_o}\right)}. \tag{10.1}$$

In this evaluation, the effective indices of the even and odd modes of the SGLC-PCF coupler are evaluated by the FVFDM [14] with PML boundary conditions [20].

10.3.1   Effect of the Structural Geometrical Parameters

The effect of the coupler geometrical parameters, rotation angle of the director of the NLC, and temperature on the coupling length of the proposed coupler is studied. The influence of the d/Λ ratio is the first parameter to be considered. Figure 10.2 shows the variation of the coupling lengths of the SGLC-PCF coupler for the two polarized modes with the d/Λ ratio at the operating wavelength of $1.55\,\mu m$. It is observed from the figure that the coupling lengths for the two polarized modes increase with increasing d/Λ at a constant hole pitch Λ of $2.0\,\mu m$. As d/Λ increases at constant Λ, the soft glass bridge between the two cores decreases. Therefore, the distance traveled by the modes to transfer from the left core to the right core, and hence the coupling lengths of the two polarized modes increase. As d/Λ increases from

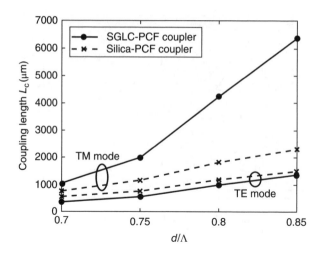

Figure 10.2 Variation of the coupling lengths of the two polarized modes with the d/Λ ratio at constant Λ of 2.0 μm. *Source*: Ref. [12]

0.7 to 0.85, the coupling lengths of the quasi-TE and quasi-TM modes of the SGLC-PCF coupler increase from 358 and 1044 μm, to 1370 and 6373 μm, respectively.

The variation of the coupling length for the two polarized modes of the conventional silica PCF coupler with air holes is also shown in Figure 10.2. In this case, the infiltrated NLC holes of diameter d_o are removed from the core regions. The refractive index of the silica material is taken as 1.45 at $\lambda = 1.55$ μm and the hole pitch is fixed to 2.0 μm. As d/Λ increases from 0.7 to 0.85, the coupling lengths of the quasi-TE and quasi-TM modes of the silica PCF coupler increase from 548 and 754 μm to 1491 and 2294 μm, respectively. It should be noted that the birefringence, defined as the difference between the effective indices of the quasi-TE and quasi-TM modes, is small for the PCF couplers with conventional silica air holes, while the SGLC-PCF coupler has high birefringence without using elliptical holes or two bigger holes in the first ring [21]. Therefore, the difference between the coupling lengths for the quasi-TE and quasi-TM modes of the SGLC-PCF coupler for a given d/Λ is greater than that of the conventional PCF coupler, as revealed in Figure 10.2.

The form birefringence [22] is defined as the ratio of $(L_{cTM} - L_{cTE})$ to L_{cTM}, where L_{cTE} and L_{cTM} are the coupling lengths of the quasi-TE and quasi-TM modes, respectively. Figure 10.3 shows the variation of the form birefringence of the SGLC-PCF coupler [12] and the PCF coupler with conventional silica air holes with d/Λ, while the hole pitch, rotation angle of the director of the NLC, temperature, and wavelength are fixed to 2.0 μm, 90°, 25°C, and 1.55 μm, respectively, and the radius of the NLC infiltrated dual cores r_o is taken as 0.5 μm. It is seen from Figure 10.3 that the form birefringence of the conventional PCF coupler and SGLC-PCF coupler increases with increasing d/Λ. As d/Λ increases from 0.7 to 0.85, the form birefringence increases from 0.657 and 0.273 to 0.796 and 0.331 for the SGLC-PCF coupler and the conventional PCF coupler, respectively. It is also evident from Figure 10.3 that the form birefringence of the SGLC-PCF coupler is approximately 2.2 times that of the conventional PCF coupler. In addition, the form birefringence values indicate that the SGLC-PCF coupler has strong polarization dependence; therefore, it can be used as a polarization splitter and its

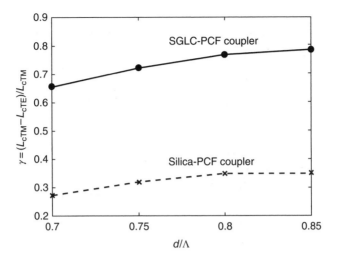

Figure 10.3 Variation of the form birefringence of the SGLC-PCF coupler and conventional silica PCF coupler with the d/Λ ratio at constant Λ of 2.0 μm. *Source*: Ref. [12]

polarization dependence is stronger than those splitters presented in Ref. [22]. However, the conventional silica PCF coupler [5] has low birefringence and high birefringence can be realized by adjusting the size of the air holes around the two cores [22–24], which increases the difference between the coupling lengths for the two polarized modes. In Ref. [23], the two identical cores are formed by a combination of large and small air holes, which makes the two cores birefringent. Zhang and Yang [24] reported a polarization splitter based on two nonidentical cores with also a combination of large and small air holes. However, in Ref. [22], two elliptic cores are used to improve the polarization dependence of the conventional silica PCF coupler.

The effect of the radius r_0 of the two identical NLC cores is also investigated. In this study, the hole pitch, d/Λ ratio, rotation angle of the director of the NLC, temperature, and wavelength are fixed to 2.0 μm, 0.8, 90°, 25°C, and 1.55 μm, respectively. Figure 10.4 shows the variation of the coupling length of the two polarized modes with the wavelength at different r_0, 0.4, 0.45, and 0.5 μm. As the wavelength increases, the confinement of the two polarized modes through the core regions decreases. Consequently, the distance traveled by the two polarized modes, and hence the coupling length, decreases on increasing the wavelength. In addition, the index contrasts seen by the two polarized modes decrease, and hence the confinement through the core regions decreases on increasing r_0. As a result, the distance traveled by the modes and then the coupling length decrease with increasing r_0.

It is also observed from Figure 10.4 that the coupling length of the quasi-TE mode at $\varphi=90°$ is shorter than that of the quasi-TM mode. At $\varphi=90°$, the relative permittivity ε_r of the E7 material has the diagonal form $[\varepsilon_{xx}, \varepsilon_{yy}, \varepsilon_{zz}]$ where $\varepsilon_{xx}=n_o^2$, $\varepsilon_{yy}=n_e^2$, and $\varepsilon_{zz}=n_o^2$. In this case, ε_{yy} is greater than ε_{xx}; therefore, the index contrast seen by the quasi-TM modes is greater than that for the quasi-TE modes. Consequently, the quasi-TM modes are more confined in the core regions than the quasi-TE modes. As a result, the quasi-TM modes travel a longer distance than the quasi-TE modes to transfer from the left core to the right core. Consequently, the coupling length of the quasi-TM mode is longer than that of the quasi-TE mode.

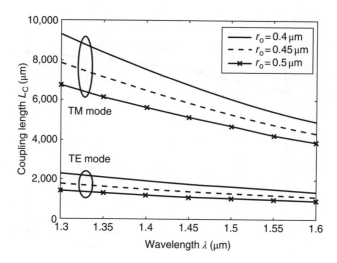

Figure 10.4 Variation of the coupling length of the two polarized modes of the SGLC-PCF coupler with the wavelength at different r_o values. *Source*: Ref. [12]

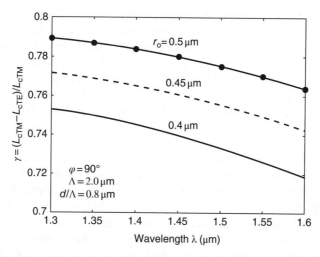

Figure 10.5 Variation of the form birefringence of the SGLC-PCF coupler with the wavelength at different r_o values. *Source*: Ref. [12]

The situation is reversed at $\varphi = 0°$ where the relative permittivity ε_r of the E7 material has the diagonal form $\left[n_e^2, n_o^2, n_o^2 \right]$. In this case, the coupling length of the quasi-TE mode is longer than that of the quasi-TM mode. Figure 10.5 shows the variation of the form birefringence with the wavelength at different r_o, 0.4, 0.45, and 0.5 µm. It is evident from this figure that the form birefringence increases with increasing r_o. At $\lambda = 1.55$ µm the form birefringences are equal to 0.7267, 0.7491, and 0.7699 at $r_o = 0.4$, 0.45, and 0.5 µm, respectively.

The effect of the deformation of the two identical NLC infiltrated holes into elliptical cores on the performance of the suggested coupler is further studied. Here, a_o and b_o are the radii of

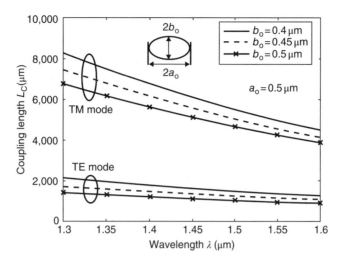

Figure 10.6   Variation of the coupling length of the two polarized modes of the SGLC-PCF coupler with the wavelength at different b_0, while a_0 is taken as 0.5 μm where a_0 and b_0 are the radii of the elliptical NLC holes in the x- and y-directions, respectively. *Source*: Ref. [12]

the elliptical holes in the x- and y-directions, respectively, as shown in the inset of Figure 10.6. The figure shows the variation of the coupling length of the two polarized modes with the wavelength at different b_0, 0.4, 0.45, and 0.5 μm. In this study, the hole pitch, d/Λ ratio, rotation angle, and temperature are fixed at 2.0 μm, 0.8, 90°, and 25°C, respectively. In addition, the radius in the x-direction a_0 is fixed at 0.5 μm. It is found that the coupling lengths of the two polarized modes decrease with increasing b_0. However, the numerical results reveal that the form birefringence increases with increasing the radius in the y-direction, b_0. At the operating wavelength of 1.55 μm, the form birefringence increases from 0.7239 to 0.7699 as b_0 increases from 0.4 to 0.5 μm. The effect of the radius a_0 in the x-direction on the performance of the reported coupler is also investigated. In this case, the hole pitch, d/Λ ratio, rotation angle, and temperature are fixed at 2.0 μm, 0.8, 90°, and 25°C, respectively. In addition, the radius in y-direction b_0 is fixed at 0.5 μm. It is found that the effect of the a_0 variation is the same as that of the b_0 variation on the performance of the SGLC-PCF coupler.

10.3.2   Effect of Temperature

It should be noted that the ordinary n_o and extraordinary n_e refractive indices of the E7 material are temperature dependent [9, 25]. Therefore, the effect of the temperature variation on the coupling length is the next parameter to be considered. Figure 10.7 shows the variation of the coupling length for the two polarized modes with the temperature at a rotation angle φ, of 90°, while the other parameters are fixed to $\Lambda = 2.0$ μm, $d/\Lambda = 0.8$, $r_o = 0.5$ μm, and $\lambda = 1.55$ μm. It can be seen from the figure that the coupling length of the quasi-TM mode decreases with increasing temperature, while the coupling length of the quasi-TE mode is nearly constant. As the temperature T increases from 15 to 45°C, the coupling length of the quasi-TM mode at $\varphi = 90°$ decreases from 4587 to 3489 μm. The dependence of the coupling length on the temperature can be explained by analyzing the dominant field components of the quasi-TE

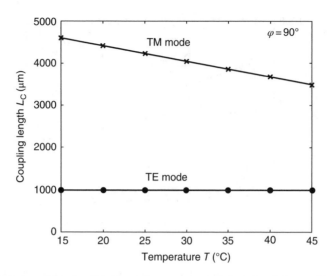

Figure 10.7 Variation of the coupling length for the two polarized modes of the dual-core SGLC-PCF coupler with the temperature. *Source*: Ref. [12]

and quasi-TM modes and the direction of the director of the NLC. At $\varphi = 90°$, the relative permittivity tensor ε_r of the E7 material has the diagonal form $[\varepsilon_{xx}, \varepsilon_{yy}, \varepsilon_{zz}]$ where $\varepsilon_{xx} = n_o^2, \varepsilon_{yy} = n_e^2$, and $\varepsilon_{zz} = n_o^2$. As the temperature increases from 15 to 45°C, ε_{yy} decreases from 2.9227 to 2.7569, while ε_{xx} changes slightly from 2.2602 to 2.2641. Therefore, the index contrast seen by the quasi-TM modes decreases with increasing temperature, while the index contrast seen by the quasi-TE modes is nearly constant. Consequently, the confinement of the quasi-TM modes inside the core regions, and hence the distance traveled by the quasi-TM modes to transfer from the left core to the right core, decreases with increasing temperature. Therefore, the coupling length of the quasi-TM mode decreases with increasing temperature, while the coupling length of the quasi-TE mode is nearly invariant, as shown in Figure 10.7.

The situation is reversed at $\varphi = 0°$ at which ε_r of the E7 material has the diagonal form $\left[n_e^2, n_o^2, n_o^2 \right]$. As the temperature increases ε_{xx} decreases, while ε_{yy} is nearly invariant. Therefore, the index contrast seen by the quasi-TE modes decreases with increasing temperature, while the index contrast seen by the quasi-TM modes is nearly constant. Consequently, the confinement of the quasi-TE modes inside the core regions and hence the distance traveled by the quasi-TE modes to transfer from the left core to the right core decreases with increasing temperature. Therefore, the coupling length of the quasi-TE mode decreases with increasing temperature, while the coupling length of the quasi-TM mode is nearly invariant. The control and tunability of the rotation angle of the NLC can be achieved with a good accuracy in a strong field limit [16] with sets of electrodes, as successfully show experimentally in Refs. [16, 26].

10.3.3 Polarization Splitter Based on PCF Coupler with LC Cores

10.3.3.1 Analysis of the Polarization Splitter

Figures 10.2 and 10.3 reveal that the SGLC-PCF coupler [12] has strong polarization dependence due to the infiltration of the NLC which increases the birefringence between the

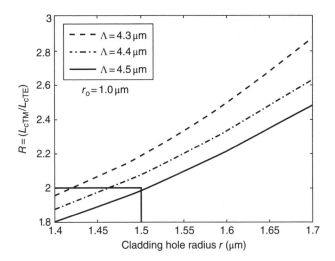

Figure 10.8 Variation of the coupling length ratio R for the quasi-TE and quasi-TM modes with the cladding hole radius r at different hole pitches. *Source*: Ref. [12]

two fundamental polarized modes. Therefore, the SGLC-PCF coupler can be easily designed as a polarization splitter and its polarization dependence is stronger than those splitters presented in Refs. [22–24]. The fiber coupler can separate the two polarized states, quasi-TE and quasi-TM modes, at a given wavelength if the coupling lengths L_{cTE} and L_{cTM} of the quasi-TE and quasi-TM modes satisfy the coupling ratio [3]:

$$R = L_{cTM} : L_{cTE} = i : j. \tag{10.2}$$

Here, i and j are two integers of different parities. In this case, the length of the coupler is equal to $L_f = L_{cTE} \times i/j$. Therefore, to achieve the shortest splitter, the optimal value of R should be 2. Figure 10.8 shows the coupling length ratio between the coupling lengths of the quasi-TE and quasi-TM modes as a function of the cladding hole radius r at different hole pitch values, 4.3, 4.4, and 4.5 µm. In this study, the central hole radius r_o, operating wavelength, rotation angle of the director of the NLC, and temperature are taken as 1.0 µm, 1.55 µm, 90°, and 25°C, respectively. It is found that the coupling length ratio R increases with increasing cladding hole radius at a given hole pitch. As can be seen from Figure 10.8, the coupling length ratio equals 1.9850 for a cladding hole radius of 1.5 µm and hole pitch of 4.5 µm. The coupling lengths calculated by the FVFDM are 3127 and 6207 µm for the quasi-TE and quasi-TM modes at the operating wavelength $\lambda = 1.55$ µm.

10.3.3.2 Beam Propagation Analysis

In order to confirm the polarization splitter based on the SGLC-PCF coupler [12], the FVFD-BPM [15] is used to study the propagation along its axial direction. Initially, at $z = 0$, the fundamental components H_y and H_x of the quasi-TE and quasi-TM modes, respectively, of

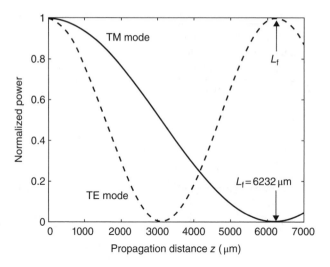

Figure 10.9 Evolution of the normalized powers at the left core for the quasi-TE and quasi-TM modes at the operating wavelength of 1.55 μm along the propagation direction. *Source*: Ref. [12]

single-core soft glass PCF with air holes obtained by the FVFDM [14] at $\lambda = 1.55\,\mu m$ are launched into the left core of the SGLC-PCF coupler. These input fields, in turn, start to transfer to the right core of the coupler and, at the corresponding coupling lengths, the fields are completely transferred to the right core. The coupling lengths calculated by the FVFD-BPM are 3128 and 6208 μm for the quasi-TE and -TM modes, respectively, which is in excellent agreement with those obtained by the FVFDM. The ratio between the coupling lengths L_{cTE} and L_{cTM} is slightly less than 2.0. Therefore, the length of the proposed splitter is $L_f = [6208 + (2 \times 3128)]/2.0 = 6232\,\mu m$ at which the two polarized states are well separated. Figure 10.9 shows the power transfer normalized to the input power for the quasi-TE and quasi-TM modes at the operating wavelength of 1.55 μm in the left core of the SGLC-PCF coupler. It is evident from Figure 10.9 that the two polarized modes are well separated after a propagation distance $L_f = 6232\,\mu m$. The normalized powers of the quasi-TE mode in the right and left cores of the coupler are 0.0005 and 0.9995, respectively, at $z = 6232\,\mu m$. However, the normalized powers of the quasi-TM mode in the right and left cores of the coupler are 0.9990 and 0.0010, respectively.

The field distributions of the dominant field component H_y and H_x of the quasi-TE and quasi-TM modes, respectively, at $\lambda = 1.55\,\mu m$ are shown in Figure 10.10 at different waveguide sections z, 0, 3128 and 6208 μm. It is evident from the figure that, at $z = 0$, the input fields are launched into the left core and, as the propagation distance increases, the normalized power in the right core increases and that in the left core decreases. At $z = 3128\,\mu m$, which is equal to the coupling length of the quasi-TE mode, the normalized power of the quasi-TE mode is almost completely transferred to the right core. The normalized powers of the quasi-TE mode in the left and right cores of the coupler are 0.0038 and 0.9962, respectively. However, the normalized powers of the quasi-TM mode in the left and right cores of the coupler are 0.4940 and 0.5056, respectively, at $z = 3128\,\mu m$. Finally, the two polarized modes are separated after a propagation distance $L_f = 6232\,\mu m$ [12].

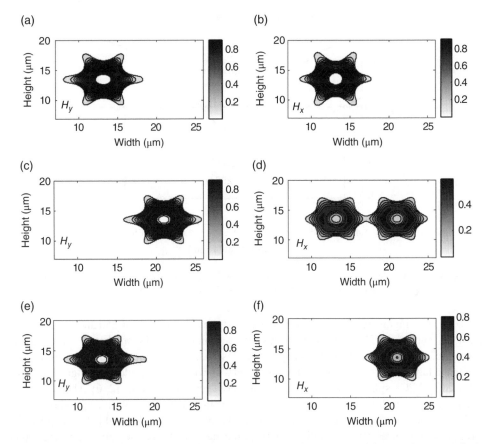

Figure 10.10   Field contour patterns for H_y and H_x of the quasi-TE and quasi-TM modes, respectively, at (a and b) $z=0$, (c and d) $z=3128\,\mu m$, and (e and f) $z=6232\,\mu m$ at $\lambda=1.55\,\mu m$. *Source*: Ref. [12]

10.3.3.3 Crosstalk

The CT is a measure of the unwanted power, remaining at the end of the SGLC-PCF coupler. The CT around the operating wavelength $\lambda=1.55\,\mu m$ for the quasi-TE and quasi-TM modes are shown in Figure 10.11. The CT in decibel [2] for the desired quasi-TE mode at the left core is defined such that

$$CT_{TE} = 10\log_{10}\left(\frac{P_{uTM}}{P_{dTE}}\right),\tag{10.3}$$

where P_{dTE} and P_{uTM} are the normalized powers of the desired quasi-TE and undesired quasi-TM modes at the left core. The CT [2] of the desired quasi-TM mode at the right core is given by

$$CT_{TM} = 10\log_{10}\left(\frac{P_{uTE}}{P_{dTM}}\right)\tag{10.4}$$

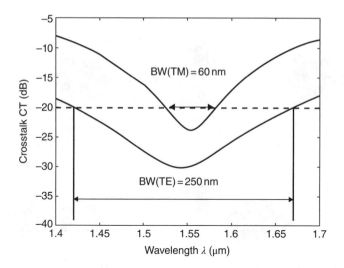

Figure 10.11 Wavelength dependence CTs of the SGLC-PCF coupler around the operating wavelength $\lambda = 1.55 \, \mu m$ for the quasi-TE and quasi-TM modes. *Source*: Ref. [12]

It is seen from Figure 10.11 that the proposed splitter has large BWs of 250 and 60 nm for the quasi-TE and quasi-TM modes, respectively, at which the CTs are better than -20 dB. Therefore, the proposed polarization splitter is less sensitive to the perturbation introduced during the fabrication process due to the low-level CTs with wide wavelength ranges. The BWs of the SGLC-PCF splitter are much larger than those reported in Refs. [2, 6]. The BW of the quasi-TE mode in Ref. [2] is 2.7 nm around $\lambda = 1.55 \, \mu m$, while the BW in Ref. [6] is 2.0 nm; further, the proposed splitter is shorter than those reported in Refs. [2] and [6] of lengths 15.4 and 9.08 mm, respectively. The SGLC-PCF splitter has a wide wavelength range, larger than the reported splitter by Chen *et al.* [3] of BW 25.4 nm around $\lambda = 1.55 \, \mu m$ for the quasi-TE mode. The splitter in Ref. [3] is also longer, 10.69 mm, than the SGLC-PCF splitter. The reported splitter is shorter than the NLC-PCF splitter [10] of coupling length 8.227 mm. Furthermore, the SGLC-PCF splitter has greater BWs than the NLC-PCF splitter [10]. Therefore, the SGLC-PCF [12] has advantages of shorter coupling length, larger BWs, and is easier to fabricate than that presented in Ref. [10].

It is also shown from Figure 10.11, that the BW of the quasi-TM mode is less than that of the quasi-TE mode. At $\varphi = 90°$, the quasi-TM mode is more confined through the core region than the quasi-TE mode. Therefore, the quasi-TE mode is more affected by the wavelength variation around $\lambda = 1.55 \, \mu m$ than the quasi-TM mode. As a result, the undesired normalized power of the quasi-TE mode at the right core at the device length of 6232 μm increases with the wavelength variation around $\lambda = 1.55 \, \mu m$ more than the undesired normalized power of the quasi-TM mode at the left core. Therefore, the BW of the quasi-TM mode at the right core is less than that of the quasi-TE mode at the left core.

10.3.3.4 Feasibility of the Polarization Splitter

The tolerances of the fiber length, rotation angle φ of the director of the NLC, and temperature are also investigated. It is worth noting that the tolerance of a specific parameter is calculated while the other parameters of the proposed design are kept constant. It is found that the fiber length and rotation angle φ allow a tolerance of $\pm 3\%$ and $\pm 5°$, respectively, at which the CTs are still better than $-20\,\text{dB}$. In addition, the CT for the quasi-TE and quasi-TM modes are better than $-20\,\text{dB}$ through the range of T from 15 to 40°C. The temperature can be controlled by using a thermo-electric module as described by Wolinski *et al.* [27, 28], allowing for temperature control in the 10–120°C range with 0.1°C long-term stability and electric field regulation in the 0–1000 V range with frequencies from 50 Hz to 2 kHz.

Finally, coupling the light in this new type of PCF polarization splitter is considered. It is found that the best way to do this is by splicing to a standard single-mode fiber (SMF) and then launching the light from a laser source direct to the SMF [29]. This approach is very effective for making a low-loss interface between SMF and PCF as experimentally reported by Leon-Saval *et al.* [29].

10.4 Summary

The coupling characteristics of a novel design of highly tunable SGLC-PCF coupler [12] have been numerically investigated. The dual-core SGLC-PCF coupler has stronger polarization dependence than the conventional silica air holes PCF coupler. In addition, a novel type of polarization splitter based on an SGLC-PCF coupler has been presented and analyzed. The SGLC-PCF splitter has advantages in terms of its short coupling length as well as low CTs over large optical BWs. The suggested splitter has a length of 6.232 mm with a CT better than $-20\,\text{dB}$ with BWs of 250 and 60 nm for the quasi-TE and quasi-TM modes, respectively. The splitter has a tolerance of $\pm 3\%$ in its length which makes the design more robust to the perturbation introduced during the fabrication. The rotation angle of the director of the NLC has a tolerance of $\pm 5°$ at which the CTs are better than $-20\,\text{dB}$. The CT for the quasi-TE and quasi-TM modes are better than $-20\,\text{dB}$ for temperatures in the range 15–40°C.

References

[1] Mangan, B.J., Knight, J.C., Birks, T.A. *et al.* (2000) Experimental study of dual core photonic crystal fiber. *Electron. Lett.*, **36**, 1358–1359.

[2] Florous, N., Saitoh, K., and Koshiba, M. (2005) A novel approach for designing photonic crystal fiber splitters with polarization-independent propagation characteristics. *Opt. Express*, **13** (19), 7365–7373.

[3] Chen, M.Y. and Zhou, J. (2006) Polarization-independent splitter based on all-solid silica-based photonic-crystal fibers. *J. Lightw. Technol.*, **24**, 5082–5086.

[4] Lægsgaard, J., Bang, O., and Bjarklev, A. (2004) Photonic crystal fiber design for broadband directional coupling. *Opt. Lett.*, **29**, 2473–2475.

[5] Saitoh, K., Sato, Y., and Koshiba, M. (2003) Coupling characteristics of dual-core photonic crystal fiber couplers. *Opt. Express*, **11**, 3188–3195.

[6] Florous, N.J., Saitoh, J.K., and Koshiba, M. (2006) Synthesis of polarization-independent splitters based on highly birefringent dual-core photonic crystal fiber platforms. *Photon. Technol. Lett.*, **18**, 1231–1233.

[7] Saitoh, K., Florous, J.N., Koshiba, M., and Skorobogatiy, M. (2005) Design of narrow band-pass filters based on the resonant-tunneling phenomenon in multi-core photonic crystal fibers. *Opt. Express*, **13**, 10327–10335.

[8] Rosa, L., Poli, F., Foroni, M. *et al.* (Feb. 2006) Polarization splitter based on a square-lattice photonic-crystal fiber. *Opt. Lett.*, **31**, 441–443.

[9] Hameed, M.F.O., Obayya, S.S.A., Al Begain, K. *et al.* (2009) Coupling characteristics of a soft glass nematic liquid crystal photonic crystal fibre coupler. *Optoelectron. IET*, **3**, 264–273.

[10] Hameed, M.F.O. and Obayya, S.S.A. (2009) Polarization splitter based on soft glass nematic liquid crystal photonic crystal fiber. *Photon. J. IEEE*, **1**, 265–276.

[11] Hameed, M.F.O., Obayya, S.S.A., and Wiltshire, R.J. (2009) Multiplexer–demultiplexer based on nematic liquid crystal photonic crystal fiber coupler. *J. Opt. Quantum Electron.*, **41**, 315–326.

[12] Hameed, M.F.O. and Obayya, S.S.A. (2011) Coupling characteristics of dual liquid crystal core soft glass photonic crystal fiber. *Quantum Electron. IEEE J.*, **47**, 1283–1290.

[13] Xiao, L., Jin, W., Demokan, M.S. *et al.* (2005) Fabrication of selective injection microstructured optical fibers with a conventional fusion splicer. *Opt. Express*, **13** (22), 9014–9022.

[14] Fallahkhair, A.B., Li, K.S., and Murphy, T.E. (2008) Vector finite difference modesolver for anisotropic dielectric waveguides. *J. Lightw. Technol.*, **26** (11), 1423–1431.

[15] Huang, W.P. and Xu, C.L. (1993) Simulation of three-dimensional optical waveguides by a full-vector beam propagation method. *IEEE J. Quantum Electron.*, **29** (10), 2639–2649.

[16] Haakestad, M.W., Alkeskjold, T.T., Nielsen, M. *et al.* (2005) Electrically tunable photonic bandgap guidance in a liquid-crystal-filled photonic crystal fiber. *IEEE Photon. Technol. Lett.*, **17** (4), 819–821.

[17] Acharya, B.R., Baldwin, K.W., Rogers, J.A. *et al.* (2002) In-fiber nematic liquid crystal optical modulator based on in-plane switching with microsecond response time. *Appl. Phys. Lett.*, **81** (27), 5243–5245.

[18] Fang, D., Yan, Q.L., and Shin, T.W. (2004) Electrically tunable liquid-crystal photonic crystal fiber. *Appl. Phys. Lett.*, **85** (12), 2181–2183.

[19] Lei, W., Alkeskjold, T.T., and Bjarklev, A. (2009) Compact design of an electrically tunable and rotatable polarizer based on a liquid crystal photonic bandgap fiber. *Photon. Technol. Lett. IEEE*, **21**, 1633–1635.

[20] Chew, W.C., Jin, J.M., and Michielssen, E. (1997) Complex coordinate stretching as a generalized absorbing boundary condition. *Microw. Opt. Technol. Lett.*, **15** (6), 363–369.

[21] Li, J., Duan, K., Wang, Y. *et al.* (2009) Design of a single-polarization single-mode photonic crystal fiber double-core coupler. *Opt. Int. J. Light Electron Opt.*, **120** (10), 490–496.

[22] Zhang, L. and Yang, C. (2004) Polarization-dependent coupling in twin-core photonic crystal fibers. *J. Lightw. Technol.*, **22** (5), 1367–1373.

[23] Zhang, L. and Yang, C. (2003) Polarization splitter based on photonic crystal fibers. *Opt. Express*, **11** (9), 1015–1020.

[24] Zhang, L. and Yang, a.C. (2004) A novel polarization splitter based on the photonic crystal fiber with nonidentical dual cores. *IEEE Photon. Technol. Lett.*, **16**, 1670–1672.

[25] Li, J., Wu, ST., Brugioni, S. *et al.* (2005) Infrared refractive indices of liquid crystals. *J. Appl. Phys.*, **97** (7), 073501–073501-5.

[26] Alkeskjold, T.T. and Bjarklev, A. (2007) Electrically controlled broadband liquid crystal photonic bandgap fiber polarimeter. *Opt. Lett.*, **32** (12), 1707–1709.

[27] Wolinski, T.R., Ertman, S., Czapla, A. *et al.* (2007) Polarization effects in photonic liquid crystal fibers. *Meas. Sci. Technol.*, **18**, 3061–3069.

[28] Wolinski, T.R., Szaniawska, K., Ertman, S. *et al.* (2006) Influence of temperature and electrical fields on propagation properties of photonic liquid crystal fibers. *Meas. Sci. Technol.*, **17** (50), 985–991.

[29] Leon-Saval, S.G., Birks, T.A., Joly, N.Y. *et al.* (2005) Splice-free interfacing of photonic crystal fibers. *Opt. Lett.*, **30**, 1629–1631.

11

Liquid Crystal Photonic Crystal Fiber Sensors

11.1 Introduction

Conventional optical fibers have been widely used for sensing applications [1]. However, the cladding region should be removed to be in close contact with the tested material, which reduces the reliability of the sensor. Recently, photonic crystal fiber (PCF) [2] has been shown to present an attractive platform for sensing applications due to its high mode confinement and flexible design. In addition, complex PCF structures can now be fabricated due to the dramatic improvement in the PCF fabrication process. PCF sensors can be used as a part of all-PCF devices so that many devices can be combined on the same PCF platform. Therefore, the integration between the different PCF devices will be straightforward.

Surface plasmon resonance (SPR) can be defined as the excitation of surface plasmon polaritons (SPPs) which are electromagnetic waves propagating at the interface between a dielectric and a metal. The SPPs are sensitive to the refractive index change of the dielectric due to the propagation along the metal/dielectric interface. Therefore, the SPR can be used for sensing applications [3], which has attracted the interest of researchers in recent years.

Recently, many PCF SPR-based sensors [4–7] have been reported and analyzed. The PCF SPR sensors depend on the excitation of the leaky core mode of the PCF and the SPP modes excited at the metal/dielectric interfaces. When the real part of the effective index of the core mode is equal to that of the SPP mode, a phase matching condition is achieved. Therefore, resonance occurs and most exciting light energy is absorbed by the surface plasmons with a sharp decrease in the incident light intensity. The matching condition can be easily tuned due to the flexibility of PCF design. In addition, the PCF sensors can be fabricated without removing the cladding region with no problem in the sensor package. In this regard, a three-hole PCF-based SPR sensor has been reported by Hautakorpi *et al.* [4]. In addition, two different PCF biomedical sensors with metallic coating have been suggested and studied by Hassani

Computational Liquid Crystal Photonics: Fundamentals, Modelling and Applications, First Edition.
Salah Obayya, Mohamed Farhat O. Hameed and Nihal F.F. Areed.
© 2016 John Wiley & Sons, Ltd. Published 2016 by John Wiley & Sons, Ltd.

and Skorobogatiy [5, 6]. However, the reported sensors [5, 6] with two large semicircular metallized channels will be difficult to fabricate using stack-and-draw technique. The metallic nanoparticles and nanowires can also be used for designing PCF SPR-based sensors. Fu *et al.* [7] have reported a PCF sensor filled with silver nanowires to achieve surface plasmon.

A PCF SPR structure with silica as a background material cannot to be used as a temperature sensor, as silica has a very low temperature coefficient. Therefore, the air holes of the silica PCF are infiltrated by liquid material with a temperature coefficient higher than that of silica. As a result, the effective index of the PCF will be affected by the temperature variation, and hence a PCF SPR temperature sensor can be realized [8]. Various materials have been adopted in temperature sensing such as water, ethanol [9], and alcohol [10]. Liquid crystal (LC) is also known for its temperature dependence [11]. However, a PCF SPR sensor with an LC as a sensing material has not yet been investigated to the best of the authors' knowledge.

In this chapter, a novel design of silica PCF SPR temperature sensor [12] with LC as the temperature-dependent material is proposed and analyzed using full-vectorial finite-element method (FVFEM) [13, 14] with perfect matched layer boundary conditions. The reported PCF design has a large central hole coated with a gold layer, which is similar to that reported in Ref. [15]. However, the central hole is infiltrated by LC of type E7. The ordinary n_o and extraordinary n_e refractive indices of the LC are temperature dependent. Therefore, the suggested design can be used as a temperature sensor. The effects of the structure geometrical parameters, LC rotation angle, and temperature are studied for structure optimization. The numerical results show that the suggested design has an average sensitivity of 0.77 nm/°C. Additionally, another temperature sensor based on plasmonic LC-PCF (PLC-PCF) [16] has been presented and analyzed. The suggested PLC-PCF can be used as a temperature sensor with tunable sensitivity of about 1 nm/°C. Further, the numerical results reveal that the reported PLC-PCF can be utilized as an ultrahigh tunable polarization filter. The reported filter has a compact device length of 0.5 mm with 600 dB/cm resonance losses at $\varphi = 90°$ for the x-polarized mode at communication wavelength of 1300 nm with low losses of 0.00751 dB/cm for the y-polarized mode. However, resonance losses of 157.71 and 0.092 dB/cm are obtained at $\varphi = 0°$ for y-polarized and x-polarized modes, respectively, at the same wavelength.

11.2   LC-PCF Temperature Sensor

11.2.1   Design Consideration

Figure 11.1a shows a cross section of the suggested LC-PCF SPR sensor [12]. The reported design is based on a silica triangular lattice PCF. The cladding air holes of diameter d are arranged with a hole pitch Λ. In addition, the first ring of the air holes is removed and a large central hole of diameter D coated with a gold layer of thickness t is inserted in the core region. The central hole is filled with nematic LC (NLC) of type E7. The ordinary n_o and extraordinary n_e refractive indices of the NLC core are given by [11]

$$n_{e,o} = A_{e,o} + B_{e,o} / \lambda^2 + C_{e,o} / \lambda^4, \tag{11.1}$$

where $A_e = 1.6933$, $B_e = 0.0078$ μm^2, $C_e = 0.0028$ μm^4, $A_o = 1.4994$, $B_o = 0.0070$ μm^2, and $C_o = 0.0004$ μm^4 at temperature $T = 25°C$. In addition, the relative permittivity tensor ε_r of the E7 material [17] is taken as

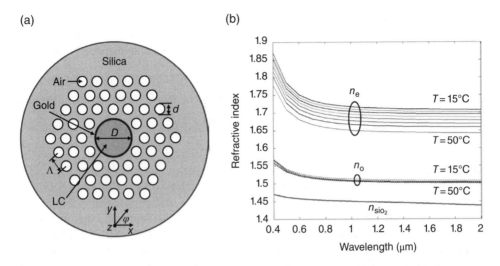

Figure 11.1 (a) Cross section of the suggested LC-PCF SPR sensor and (b) the wavelength-dependent refractive indices of the silica and n_o and n_e of the NLC material

$$\varepsilon_r = \begin{pmatrix} n_o^2 \sin^2 \varphi + n_e^2 \cos^2 \varphi & \left(n_e^2 - n_o^2\right)\cos\varphi \sin\varphi & 0 \\ \left(n_e^2 - n_o^2\right)\cos\varphi \sin\varphi & n_o^2 \cos^2 \varphi + n_e^2 \sin^2 \varphi & 0 \\ 0 & 0 & n_o^2 \end{pmatrix}, \tag{11.2}$$

where φ is the rotation angle of the director n of the NLC shown in Figure 11.1a. In this study, in-plane alignment, control, and tunability of the rotation angle of the NLC can be obtained with high accuracy using sets of electrodes and a strong field limit [18, 19].

The wavelength-dependent refractive index n_{SiO_2} of the silica background is calculated from [20]

$$n_{SiO_2} = \sqrt{1 + \frac{B_1 \lambda^2}{\lambda^2 - C_1} + \frac{B_2 \lambda^2}{\lambda^2 - C_2} + \frac{B_3 \lambda^2}{\lambda^2 - C_3}}, \tag{11.3}$$

where λ is the wavelength in μm, B_1, B_2, and B_3 are given by 0.6961663, 0.4079426, and 0.8974794, whilst C_1, C_2, and C_3 are equal to 0.00467914826, 0.0135120631, and 97.9340025 μm^2, respectively. Figure 11.1b shows the wavelength dependence of the silica, n_o and n_e of the NLC at different temperatures [11]. It is revealed from the figure that n_e is more dependent on the temperature than n_o. In addition, n_o and n_e are higher than the refractive index of the silica background material. Therefore, the propagation occurs through the suggested design by a modified total internal reflection mechanism.

The gold layer has a relative permittivity of [21]

$$\varepsilon_{Au}(\omega) = \varepsilon_\infty - \frac{\omega_p^2}{\omega(\omega + j\omega_t)}, \tag{11.4}$$

where ε_∞ is the relative permittivity at infinite frequency, ω_p is the metal plasma frequency, and ω_t is the collision frequency. For gold, ε_∞, ω_p, and ω_t are taken as 9.75, 1.36×10^{16}, and 1.45×10^{14}, respectively. The confinement loss in dB/m is calculated using the imaginary part of the effective index of the mode as follows:

$$\alpha\,(\mathrm{dB/m}) = \frac{40\pi\,\mathrm{Im}\left(n_{\mathrm{eff}}\right)}{\ln(10)\lambda}.$$

(11.5)

It should be noted that the quasi-transverse electric (TE) mode is more affected by ε_{xx} of the relative permittivity tensor ε_r of the E7 material, while the quasi-transverse magnetic (TM) mode relies on ε_{yy}. At $\varphi = 0°$, ε_r of the NLC is the diagonal of $\left[n_e^2, n_o^2, n_o^2 \right]$. The n_e of the E7 material decreases from 1.7096 to 1.6438 on increasing the temperature from 15 to 50°C at $\lambda = 1.55$ μm. However, n_o decreases slightly from 1.5034 to 1.5017 on increasing the temperature from 15 to 35°C [11], and then increases from 1.5017 to 1.5089 when T increases from 35 to 50°C at $\lambda = 1.55$ μm. Therefore, ε_{xx} is more affected by the temperature variation, while ε_{yy} is nearly invariant. Consequently, the quasi-TE modes are more affected by temperature variation than the quasi-TM modes at $\varphi = 0°$. Figure 11.2 shows the loss spectra of the quasi-TE core-guided mode at $\varphi = 0°$, and $T = 25°$C. In this study, the hole pitch Λ, d/Λ ratio, core diameter D, and metal thickness are fixed to 2.0 μm, 0.6, 2.6 Λ, and 40 nm, respectively. It is evident from the figure that there are three major attenuation peaks for the core-guided mode with 331.57, 687.40, and 2195.21 dB/cm at $\lambda = 544$, 890, and 1073 nm, respectively. The detected three peaks are due to the excitation of plasmonic modes on the surface of the metal layer that surrounds the NLC core at $T = 25°$C. However, the numerical results reveal that the first resonance peak is more sensitive to the temperature variation. Therefore, it will be studied in the subsequent simulations in this chapter.

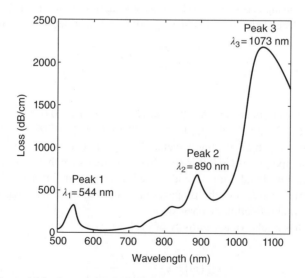

Figure 11.2   Loss spectra of the PCF core-guided mode at $T = 20°$C

11.2.2   Effects of the Structural Geometrical Parameters

The effects of the structure geometrical parameters of the proposed LC-PCF SPR on the sensor performance are investigated to achieve high spectral sensitivity. The surface plasmon waves are very sensitive to the thickness of the metal layer. Therefore, the effect of the metal layer thickness on the loss spectra is first investigated. In this study, the other parameters are fixed to $\varphi = 0°$, $T = 20°C$, $\Lambda = 2.0$ μm, $d/\Lambda = 0.6$, and $D = 2.6\ \Lambda$. Figure 11.3 shows the loss spectra of the quasi-TE core-guided mode of the suggested design at different metal thicknesses, 40, 50, and 60 nm. As the metal thickness is reduced, the SPP modes excited on both sides of the metal start coupling with each other, resulting in mixed modes. Therefore, the confinement loss increases by decreasing the metal thickness, as shown in Figure 11.3. In addition, the resonant wavelength is shifted toward longer wavelength by increasing the metal thickness. As the metal thickness increases from 40 to 60 nm, the resonant wavelength increases from 542 to 562 nm.

Figure 11.4 shows the variation of the real part of the effective indices of the fundamental quasi-TE core mode H_y^{11}; the higher order modes, H_y^{02} and H_y^{20}; and the first three surface plasmon modes, SP^1, SP^2, and SP^3. In addition, the inset figures show E_x field profiles of each mode at $\lambda = 542$ nm. As shown from the figure, the effective index of each of the first three core modes crosses that of the three lower order surface plasmon modes. Therefore, a phase matching condition is achieved for each core mode at three different wavelengths, $\lambda = 532$, 535 and 540 nm for the fundamental core mode, $\lambda = 533$, 537, and 542 nm for the H_y^{02} higher order core mode, and $\lambda = 534$, 538, and 543 nm for the H_y^{20} higher order core mode. It is noted that the wavelength values at which phase matching is achieved for the three core modes are very close and hence, when the coupling between one core mode and SP mode is strong enough, no other mode coupling occurs at this wavelength. Subsequently, only one peak appears in the loss spectrum of each core mode. Therefore, there are no other peaks in the loss curve given in Figure 11.3.

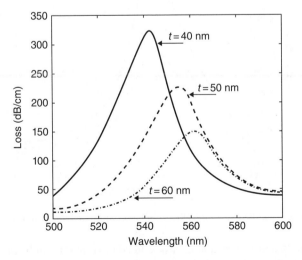

Figure 11.3   Confinement loss spectra of the LC-PCF core-guided mode at different metal thicknesses

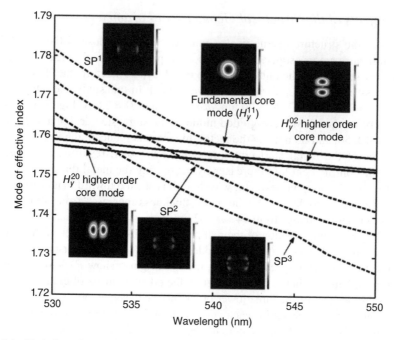

Figure 11.4 Variation of the real part of the effective indices of the core-guided mode and the first three surface plasmon modes

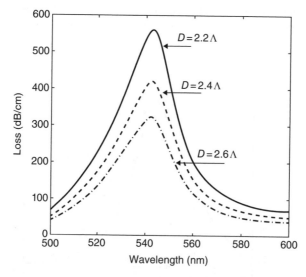

Figure 11.5 Confinement loss spectra of the LC-PCF core-guided mode at different core diameters

The influence of the NLC core diameter is studied next. In this study, the other parameters are taken as $\varphi = 0°$, $T = 20°C$, $\Lambda = 2.0$ μm, and $d/\Lambda = 0.6$. Figure 11.5 shows the loss spectra of the quasi-TE core-guided mode at different core diameters, 2.2, 2.4, and 2.6 Λ. It is evident from the figure that the confinement loss decreases on increasing the core diameter due to the

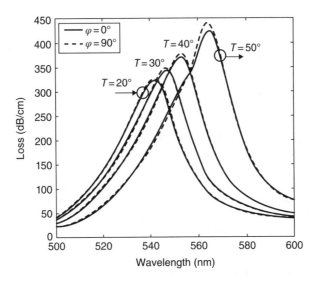

Figure 11.6 Confinement loss spectra of the quasi-TE mode at $\varphi = 0°$ and the quasi-TM mode at $\varphi = 90°$ of the LC-PCF core-guided mode at different temperatures

good confinement of the core-guided mode through the core region. As the core diameter increases from 2.2 to 2.6 Λ, the resonant wavelength slightly increases from 542 to 544 nm.

11.2.3   Effect of the Temperature

The impact of the temperature on the performance of the LC-PCF SPR sensor is also reported. Figure 11.6 shows the loss spectra of the quasi-TE core mode at $\varphi = 0°$, and the quasi-TM core mode at $\varphi = 90°$ at different temperatures, 20, 30, 40, and 50°C. In this investigation, the other parameters are fixed to $\Lambda = 2.0$ μm, $d/\Lambda = 0.6$, $D = 2.6 \Lambda$, and $t = 40$ nm. At $\varphi = 0°$, ε_r of the E7 material is the diagonal of $\left[n_e^2, n_o^2, n_o^2 \right]$, while ε_r is equal to the diagonal of $\left[n_o^2, n_e^2, n_o^2 \right]$ at $\varphi = 90°$. Therefore, ε_{xx} at $\varphi = 0°$ is equal to ε_{yy} at $\varphi = 90°$. Consequently, the modal properties and hence the confinement loss of the quasi-TE core mode at $\varphi = 0°$ are approximately equal to those of the quasi-TM mode at $\varphi = 90°$, as shown in Figure 11.6. It is also revealed from the figure that the shift in the resonance wavelength increases with increasing temperature. The resonance wavelengths of the quasi-TE modes at $\varphi = 0°$ are equal to 542, 547, 553, 565 nm, at $T = 20$, 30, 40, and 50°C, respectively.

11.2.4   Effect of the LC Rotation Angle

The rotation angle of the NLC effect on the performance of the reported sensor is introduced next, whilst the other parameters are fixed to $T = 20°C$, $\Lambda = 2.0$ μm, $d/\Lambda = 0.6$, and $D = 2.6 \Lambda$. The numerical results show that the rotation angle has no effect on the resonance wavelength. In the rotation angle range $0° \leq \varphi \leq 45°$, ε_{xx} is more dependent on n_e of the NLC material while ε_{yy} is dependent on n_o. Therefore, the quasi-TE modes are temperature dependent, while the quasi-TM modes are not. However, ε_{xx} and ε_{yy} are dependent on n_o and n_e, respectively, in the rotation angle φ range greater than 45–90°. As a result, the quasi-TM modes are temperature dependent, while the quasi-TE modes are not for $45° < \varphi \leq 90°$.

11.2.5   Sensitivity Analysis

The sensitivity of the proposed LC-PCF SRR sensor is finally calculated using the wavelength interrogation method as follows:

$$S_\lambda = \frac{\Delta\lambda_p}{\Delta T} \, \text{nm} \, / \, °\text{C}. \tag{11.6}$$

Here, Δe_p represents the peak position shift of the resonant core mode loss for ΔT change of the NLC temperature. Using the numerical results shown in Figure 11.6, $\Delta\lambda_p$ is equal to 5, 6, and 12 nm corresponding to temperature intervals [20–30°C], [30–40°C], and [40–50°C], and hence the sensitivity of 0.5, 0.6, and 1.2 nm/°C, respectively, can be obtained. Therefore, the proposed sensor has an average sensitivity of 0.77 nm/°C, which is superior to the sensitivity of the temperature sensors of [10] with ethanol that achieved 0.35 nm/°C and to that of Ref. [8] with 0.2 nm/°C with almost the same fabrication complexity.

11.3   Design of Single Core PLC-PCF

In this section, a novel design of a single-core PLC-PCF [16] is proposed and analyzed using the FVFEM [22] with a minimum element size of 0.0008 μm. The PLC-PCF design has a metal wire at the central hole of the cladding region. In addition, a large hole in the core region is infiltrated with an NLC of type E7. The effects of the structure geometrical parameters, temperature, and rotation angle of the director of the NLC on the modal characteristics of the proposed design are investigated in detail. The analyzed parameters are the effective index (n_{eff}) and the attenuation loss (α). It is evident from the simulation results that the reported design has high tunability with external electric field applied to the NLC infiltrated in the fiber core. Additionally, the PLC-PCF can be used as a temperature sensor with tunable sensitivity of about 1 nm/°C. Moreover, the suggested design can be operated as a filter of compact device length of 0.5 mm that can pass the y-polarized mode at $\varphi = 90°$ with high losses of 600 dB/cm for the x-polarized mode at the communication wavelength $\lambda = 1300$ nm. However, the x-polarized mode can be obtained at the same wavelength at $\varphi = 0°$ with resonance losses of 157.71 dB/cm for the y-polarized mode. Therefore, the suggested filter has higher tunability due to the infiltration of the NLC material than those reported in the literature [23, 24]. The suggested filter has better polarization characteristics than that investigated by Nagasaki *et al.* [23]. The resonance strength of the PCF selectively filled with metal wires into cladding air holes is not strong enough and there is more than one resonance peak [23]. Additionally, the proposed PLC-PCF filter has stronger resonance strength than that presented by Xue *et al.* [24] for the same communication window. Further, the reported structure in Ref. [24] depends on using PCF with different hole diameters in the cladding region. Moreover, the suggested filter has high tunability due to the infiltration of NLC material. In the following subsections, the performance of the PLC-PCF design as a filter as well as a temperature sensor will be explained thoroughly.

11.3.1   Design Consideration

Figure 11.7 shows the proposed single-core PLC-PCF[16]. The suggested PCF has a fluorite crown of type FK51A as a background material, and the cladding air holes of diameter

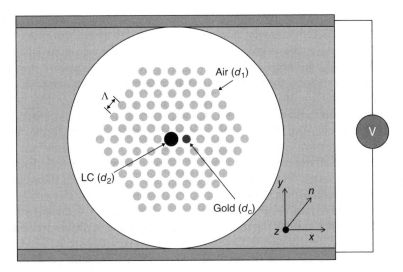

Figure 11.7 Cross section of the PLC-PCF filter filled with a metal wire and sandwiched between two electrodes. *Source*: Ref. [16]

$d_1 = 2\,\mu$m are arranged in a hexagonal shape with a lattice constant $\Lambda = 3.75\,\mu$m. In addition, a gold metal wire of diameter $d_c = 2\,\mu$m is selectively filled in the cladding air holes. Moreover, a large hole of diameter $d_2 = 3.4\,\mu$m infiltrated by NLC is inserted in the core region. The relative permittivity tensor of the NLC takes the same form as Eq. (11.2).

The Sellmeier equation of the FK51A material [25] is given by

$$n^2\left(\lambda\right) = 1 + \frac{A_1\lambda^2}{\lambda^2 - B_1} + \frac{A_2\lambda^2}{\lambda^2 - B_2} + \frac{A_3\lambda^2}{\lambda^2 - B_3},$$

where $n(\lambda)$ is the wavelength-dependent refractive index of the FK51A, $A_1 = 0.971247817$, $A_2 = 0.2169014$, $A_3 = 0.9046517$, $B_1 = 0.00472302\,\mu$m^2, $B_2 = 0.01535756\,\mu$m^2, and $B_3 = 168.68133\,\mu$m^2. The proposed design depends on a high birefringent NLC core and a metal wire filled in one of the cladding air holes. The NLC has two principal refractive indices: n_e and n_o for the extraordinary and ordinary rays, respectively. The coupling between the two orthogonal polarized core modes (x- and y-polarized modes) and the plasmonic modes can be controlled using the birefringence of the NLC material. The energy coupling efficiency between the plasmonic modes and one of the orthogonal polarization core modes (the mode to be suppressed) can be enhanced by decreasing the difference between the ordinary refractive index of the NLC core $\left(n_o = 1.5037$ at $\lambda = 1.3\,\mu$m$\right)$ and that of the background material. Through the reduction of the index difference, the coupling between the lossy-in-nature plasmonic modes and the polarization mode (to be suppressed) can be highly improved; consequently, the core mode (to be suppressed) starts to lose power very quickly, leading to a more compact mode filter design. Hence, the choice of the FK51A glass $\left(n = 1.4777$ at $\lambda = 1.3\,\mu$m$\right)$ is justified to give us a better control over the offered design in terms of compactness. Additionally, optical fiber based on FK51A glass can now be fabricated using the well-known stack-and-draw technology [26]. Therefore, the suggested FK51APCF can be fabricated using the stack-and-draw technique [26].

The relative permittivity of gold in the visible and near IR-region can be expressed by Eq. (11.4) [27]. It is worth knowing that the effective index of the NLC material at any wavelength through the range is greater than that of the background material at the same wavelength. Therefore, the propagation through the NLC core occurs by modified total internal reflection. In addition, there are two main fundamental core modes that will be guided through the fiber core, x-polarized and y-polarized modes. One of these two fundamental modes will couple with the surface of the metal wire, while the other will propagate through the fiber without any coupling. The coupling operation depends on the molecule directions of the NLC material which will determine the coupled core mode to the metal surface. The n_e and n_o of the E7 material are temperature dependent. Consequently, the reported filter has high tunability with temperature. In addition, the performance of the reported filter can be tuned using the structure geometrical parameters.

First, numerical investigation of the electric field distribution across the suggested design was carried out. In order to study the strength and the distribution of the electric field through the proposed structure, the equation $\nabla \cdot D = \varepsilon(x,y)E = 0$ was solved using the FEM where D is the dielectric displacement, E is the electric field, and $\varepsilon(x,y)$ is the dielectric permittivity profile. In the numerical simulation, to reduce the simulation time, the width of the electrodes was taken as 12 times the hole pitch of the proposed structure, as shown in Figure 11.8. This was enough to avoid electrode edge effects that could influence the field within the fiber. The polarization extinction ratio for LC is continuously tuned when the driving voltage V is above the Fredericks threshold [28]. Therefore, in this simulation $V_{rms} = 40$ V is applied. Figure 11.8 shows the contour lines of the electric potential (V) (horizontal gray solid lines) and normalized

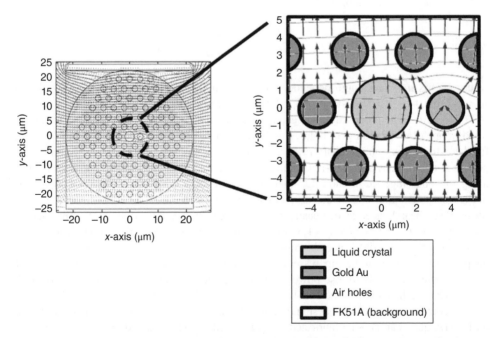

Figure 11.8 Contour lines of electric potential (V) (horizontal gray solid lines) and normalized arrow surface of E-field distribution (vertical arrows) across the proposed structure. *Source*: Ref. [16]

arrow surface of the E-field distribution (vertical arrows) across the proposed structure. It is revealed from the figure that the E-field distribution is still uniform inside the NLC core even in the presence of metal wires. Moreover, although, expected to be nonuniform around the metal wire, the electric field lines are clearly uniform inside the NLC core. As a result, an accurate director angle $\varphi = 0°$ or $\varphi = 90°$ can be obtained using the two electrodes arrangement.

The polarization characteristics of the two fundamental polarized modes of the suggested filter are then studied and analyzed. Figure 11.9 shows the wavelength dependence of the

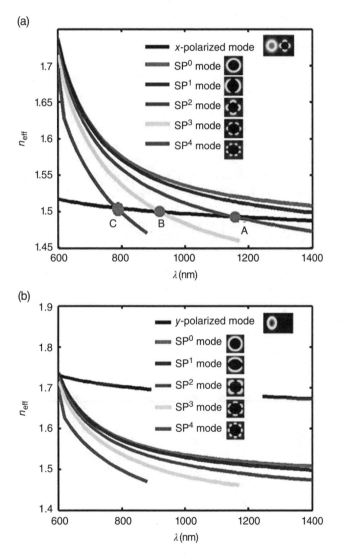

Figure 11.9 Wavelength dependence of the effective indices of (a) the x-polarized core mode and surface plasmon SP⁰, SP¹, SP², SP³, and SP⁴) modes. (b) The y-polarized core mode and surface plasmon (SP⁰, SP¹, SP², SP³, and SP⁴) modes. Points A, B, and C are the intersection points of the x-polarized core mode dispersion curve with the SP², SP³, and SP⁴ modes dispersion curves, respectively, while $T = 25°C$ and $\varphi = 90°$.
Source: Ref. [16]

effective indices of the different modes in the PCF (the x-polarized core mode, the y-polarized core mode, SP^0, SP^1, SP^2, SP^3, and SP^4 modes) at $T = 25\,^\circ C$ and $\varphi = 90^\circ$. At $\varphi = 90^\circ$, the director of the NLC is normal to E_x and is parallel to E_y and the dielectric permittivity tensor ε_r of the E7 material has the diagonal form $\left[n_o^2, n_e^2, n_o^2 \right]$. In this case, ε_{xx} is smaller than ε_{yy}; therefore, the effective index of the x-polarized mode is smaller than that of the y-polarized mode, as shown in Figure 11.9. It is also seen from Figure 11.9b that the effective index of the y-polarized mode is greater than the SP modes; therefore, no coupling occurs.

It is also evident from Figure 11.9 that the x-polarized mode is only coupled to the higher order surface plasmon modes (SP^2, SP^3, and SP^4) due to phase matching between them at resonance wavelengths of 1162, 920, and 785 nm, respectively. In this case, the effective indices for the two coupled modes are equal at their resonance wavelengths. However the y-polarized core mode makes no coupling because of the absence of phase matching with the surface plasmon modes. Therefore, the polarization filter behavior is very obvious as the y-polarized core mode will pass through the fiber with no change, while the x-polarized core mode will not pass due to the coupling with the surface plasmon modes.

Figure 11.10 shows the calculated attenuation loss of the two fundamental core modes. It can be seen that there are three peaks at $\lambda = 1162$, 920, and 780 nm due to the coupling of the x-polarized core mode with the SP^2, SP^3, and SP^4 modes, respectively. The three coupling points with SP^2, SP^3, and SP^4 modes are indicated in Figure 11.10 as points A, B, and C, respectively. On the other hand, the attenuation losses of the y-polarized mode are very low for the entire wavelength range, which reveals the behavior of the polarization filter.

In order to prove our results, field plots of the two polarized core modes at different wavelengths 900, 1162 and 1300 nm are shown in Figure 11.11. As may be seen, there is a strong coupling between the x-polarized core mode and the SP^2 mode at the resonance wavelength $\left(\lambda_r = 1162\,\text{nm} \right)$. However, at $\lambda = 900\,\text{nm}$ and $\lambda = 1300\,\text{nm}$, most of the electric field is coupled to the fiber core. It is also clear from Figure 11.11 that the y-polarized mode is well confined in the fiber core through the entire wavelength range with no SP modes coupling.

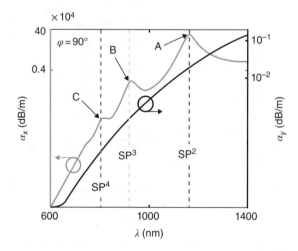

Figure 11.10 Loss spectrum (in log scale) for the x-polarized and y-polarized core modes at $T = 25\,^\circ C$, and $\varphi = 90^\circ$. Points A, B, and C are the equivalent points to the resonance wavelengths shown in Fig. 11.9a. *Source*: Ref. [16]

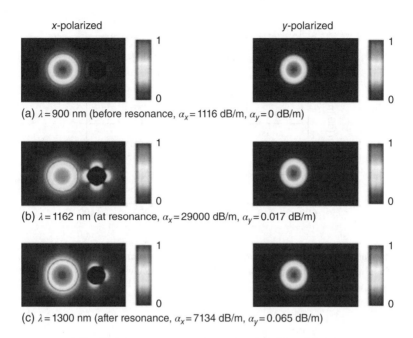

Figure 11.11 Field plots of x- and y-polarized core modes at different wavelengths: (a) before resonance at $\lambda = 900$ nm, (b) at resonance at $\lambda = 1162$ nm, and (c) after resonance at $\lambda = 1300$ nm. *Source*: Ref. [16]

11.3.2   Effect of the LC Rotation Angle

One of the most effective parameters on the behavior of the suggested filter is the rotation angle (φ) of the NLC material which can be easily controlled by changing the applied electric field. Figure 11.12 shows the attenuation loss behavior of the two core modes: x-polarized and y-polarized modes at $\varphi = 0°$ and $\varphi = 90°$. It is clear that the filter operation is inverted with the rotation angle change. At $\varphi = 0°$, the y-polarized mode couples with the SP modes, while the x-polarized mode makes no coupling. It should be noted that the x-polarized mode is coupled strongly at $\varphi = 90°$ with a loss peak of $\alpha_x = 28{,}500$ dB/m, which is larger than the other peak of the coupled y-polarized mode at $\varphi = 0°$ with $\alpha_y = 8{,}500$ dB/m. This is due to the position of the metal wire, which is aligned horizontally to the core region.

At $\varphi = 0°$, the permittivity tensor ε_r of the NLC has the diagonal form $\left[n_e^2, n_o^2, n_o^2 \right]$. Therefore, ε_{xx} is greater than ε_{yy}, and hence the x-polarized mode effective index is greater than those of the SP modes. Therefore, no coupling occurs with the x-polarized mode at $\varphi = 0°$. It is also evident from Figure 11.12 that the rotation angle has no effect on the resonance wavelength at which the two polarized core modes are coupled to the SP modes at $\varphi = 0°$ and $\varphi = 90°$.

11.3.3   Effect of the Structural Geometrical Parameters

The polarization characteristics of the designed filter can be altered by changing the geometrical parameters of the PCF. In this part, the impacts of the diameter of the cladding air holes (d_1), the metal diameter (d_c), and the NLC central hole diameter (d_2) are analyzed and investigated.

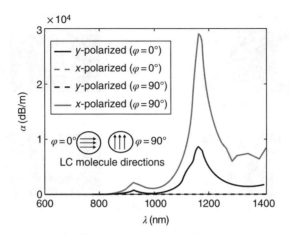

Figure 11.12 Variation of the attenuation losses of the two fundamental polarized modes at $\varphi = 90°$ and $\varphi = 0°$ when the temperature is fixed at $T = 25°C$. The molecules directions relative to the rotation angle are shown in the inset. *Source*: Ref. [16]

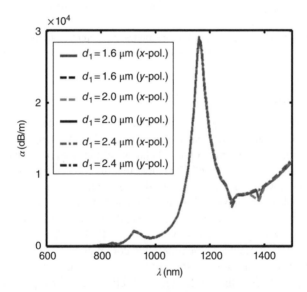

Figure 11.13 Variation of the wavelength-dependent attenuation losses of the two fundamental core modes (x-polarized and y-polarized modes) at different air hole diameters (d_1), 1.6, 2, and 2.4 μm while $d_c = 2$ μm, $d_2 = 3.4$ μm, $T = 25°C$, and $\varphi = 90°$. *Source*: Ref. [16]

Figure 11.13 shows the variation of the two fundamental core-guided modes' attenuation losses at different diameters of the cladding air holes (d_1). It can be seen that d_1 has no effect on the resonance wavelength of the x-polarized mode as the mode field is well confined through the core region. It is worth noting that the y-polarized mode has approximately no losses over the whole spectrum.

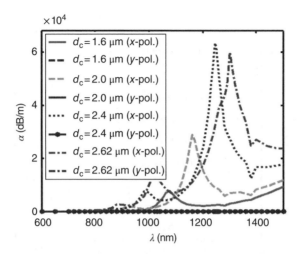

Figure 11.14 Variation of attenuation losses of the two fundamental core modes (*x*-polarized and *y*-polarized modes) with the metal wire diameter (d_c), while d_1, d_2, *T*, and φ are taken as 2 μm, 3.4 μm, 25 °C, and 90°, respectively. *Source*: Ref. [16]

It is well known that surface plasmon waves are extremely sensitive to the metal thickness. Figure 11.14 shows the variation of the attenuation loss of the two polarized modes as a function of the metal wire diameter (d_c), while all other PCF parameters are fixed at their original values. It can be seen that the resonance wavelength increases as the metal wire diameter increases and the loss strength is significantly affected by changing the metal wire diameter. Therefore, one needs to be careful when choosing the metal wire diameter so that the desired effects can be achieved. It should be noted that at $d_c = 2.62$ μm, the resonance wavelength is approximately equal to 1,300 nm with resonance losses of 60,000 and 0.751 dB/m at $\varphi = 90°$ for *x*- and *y*-polarized modes, respectively. However, resonance losses of 15,771 dB/m at $\varphi = 0°$ can be achieved for the *y*-polarized mode at the same wavelength with 9.2 dB/m losses for the *x*-polarized mode. It is well known that the wavelength of 1300 nm lies in the second communication window of optical fibers $(1250 – 1350)$ nm. For a filter design, it is required to achieve a compact device length with high losses for one polarized mode and very low losses for the other. Therefore, in order to optimally design the filter, it is required to achieve total leakage losses for the *y*-polarized mode of <0.1 dB, while the total leakage losses for the suppressed *x*-polarized mode are > 30 dB. It is found that these requirements can be achieved with a compact device length of 0.5 mm at $\varphi = 90°$.

However, at $\varphi = 0°$ a compact device length of 1.9 mm will be needed to pass the *x*-polarized only. Therefore, a suitable electrode size can be chosen to cover the PCF length to keep the same refractive indices in the NLC core. It should also be noted that similar to any other fiber types, bending with small radii of curvature causes radiation losses. However, in the suggested design and thanks to the excellent confinement of light, the proposed fiber is expected to be subject to reduced radiation losses from the bending. Additionally, the compact length can reduce the bending of the suggested filter.

Figure 11.15 shows the effect of variation of the attenuation loss of the two fundamental core modes (*x*-polarized and *y*-polarized) with the diameter (d_2) of the NLC infiltrated hole in

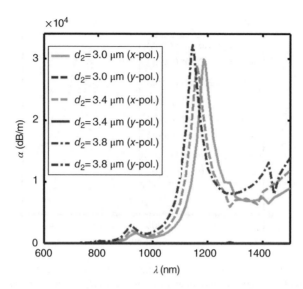

Figure 11.15 Variation of attenuation losses of the two fundamental core modes (x-polarized and y-polarized modes) with the NLC central hole diameter (d_2), while d_1, T, and φ are taken as 2 μm, 25°C, and 90°, respectively. *Source*: Ref. [16]

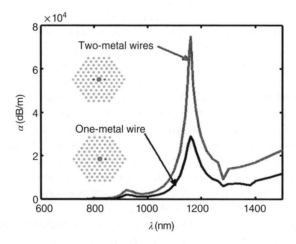

Figure 11.16 Variation of the attenuation losses of the x-polarized core mode of the PLC-PCF with one and two metal wires with the wavelength. *Source*: Ref. [16]

the fiber core. It is seen that the diameter d_2 has a slight effect on the resonance wavelength. As d_2 increases from 3 to 3.8 μm, the resonance wavelength decreases from 1185 to 1145 nm.

A study on the effect of the number of metal wires surrounding the NLC core is also investigated and analyzed. Figure 11.16 shows the variation of the wavelength dependent attenuation losses of the x-polarized core mode of the PLC-PCF with one and two metal wires. It is seen that using metal wires on the two sides of the NLC core will strongly enhance the filter behavior. The attenuation loss of the PLC-PCF with two metal wires $(\alpha_x = 75,000 \text{ dB/m})$ at the resonance wavelength will be three times larger than that with only one metal wire $(\alpha_x = 28,500 \text{ dB/m})$. It is also clear from Figure 11.16 that the two cases have the same resonance wavelength. However, using a second metal wire results in a sharper attenuation curve. Therefore, it is expected that increasing the number of metal wires around the NLC core will strongly enhance the filter operation.

It should be noted that the absorption and scattering losses of the NLC are much larger than those of the background materials, such as silica, Pyrex, and FK51A [29]. Therefore, the background material loss can be ignored. In addition, the scattering losses of the bulk NLC of around 15–40 dB/cm reported by Hu and Whinnery [30] are much greater than its absorption losses at visible and near-infrared wavelengths. However, Green and Madden [31] show that the scattering loss of NLCs can be decreased from 1 to 3 dB/cm by infiltrating the LC into small capillaries with inner diameters of 2–8 µm. In this study [31], conventional Pyrex fibers with refractive index of 1.470 and core diameter of 4 µm were infiltrated by NLC whose director was aligned along the fiber axis. Laser light of wavelength 633 nm was focused on the fiber end and the scattered light out of the fiber normal to the propagation direction was observed. In this case, an average loss of 3.25 dB/cm was obtained. The scattering losses of silica fiber infiltrated by NLC have also been investigated [31]. Scattering losses of 1.5–2.4 dB/cm were obtained along a fiber length of 30 cm. Therefore, the scattering losses in small core fibers are 20 dB/cm obtained after propagation in a slab waveguide or in a large diameter LC cored fiber [30]. Moreover, the scattering loss of the LC is dependent on the filling technique. In this regard, the infiltration of the LC using capillary forces has lower losses than when using high pressure [31].

The suggested PLC-PCF has a central hole of diameter $d_2 = 3.4$ µm which is infiltrated by the NLC where the light is propagating through the NLC core in a way similar to Ref. [31]. The reported filter has a compact length of 0.5 mm. Further, a fluorite crown of type FK51A of refractive index of 1.477 is used as a background material, which has a refractive index value very close to Pyrex fibers with refractive index of 1.470 [31]. Therefore, it is believed that the scattering losses will be less than 1.5–2.4 dB/cm [31], giving rise to low scattering losses of 0.075–0.2 dB for our device of compact length of 0.5 mm. However, for a longer device where scattering losses can be an issue, an alignment layer can be used at the edges to help align the molecules near the edges so the scattering losses can be minimized.

11.3.4   Effect of the Temperature

The effect of the temperature on the performance of the proposed filter is also investigated. Figure 11.17 shows the effect of variation of temperature on the filter behavior and resonance wavelength at $\varphi = 90°$. It is seen that the resonance wavelength of the coupled core-guided mode has a tunable behavior with the temperature. In this study, four different cases have been reported and discussed at $T = 15, 25, 35,$ and $45°C$. The resonance wavelengths are found to be 1150, 1160, 1170, and 1140 nm, respectively. From these results, we can conclude that a 10°C change in temperature gives a shift of 10 nm in the resonance wavelength (λ_r). Therefore, the proposed polarization design can be used as temperature sensor with tunable sensitivity of about 1 nm/°C.

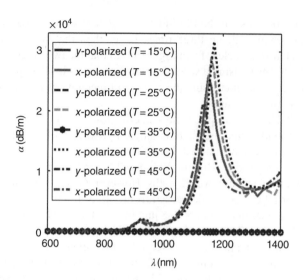

Figure 11.17 Variation of attenuation losses of the two fundamental polarized core modes at different temperatures, while the rotation angle φ is fixed at 90°. *Source*: Ref. [16]

11.4 Summary

In this chapter, a novel deign of LC-PCF SPR temperature sensor is proposed and studied. The suggested sensor has an average sensitivity of 0.77 nm/°C, which is comparable to the sensitivity of the temperature sensors recently reported. Additionally, another PLC-PCF has been introduced and analyzed. The reported PLC-PCF can be used as temperature sensor with tunable sensitivity of about 1 nm/°C. Numerical results demonstrate that the proposed PLC-PCF can be used as ultrahigh tunable polarization filter. The suggested design has a central hole infiltrated with an NLC that offers high tunability with temperature and external electric field. The results show that the resonance losses and wavelengths are different in the x- and y-polarized directions depending on the rotation angle φ of the NLC. The reported filter of compact device length 0.5 mm can achieve 600 dB/cm resonance losses at $\varphi = 90°$ for the x-polarized mode at a communication wavelength of 1300 nm with low losses of 0.00751 dB/cm for the y-polarized mode. However, resonance losses of 157.71 dB/cm at $\varphi = 0°$ can be achieved for the y-polarized mode at the same wavelength with low losses of 0.092 dB/cm for the x-polarized mode. This means that the polarizer behavior can be inverted by changing the rotation angle of the NLC from $\varphi = 90°$ to $\varphi = 0°$. Also, the performance of the PLC-PCF filter can be enhanced by increasing the number of metal nanowires used.

References

[1] Grattan, K.T.V. and Sun, T. (2000) Fiber optic sensor technology: an overview. *Sens. Actuators A: Phys.*, **82**, 40–61.

[2] Russell, P.S.J. (2003) Photonic crystal fibers. *Science*, **299**, 358–362.

[3] Homola, J., Yee, S.S., and Gauglitz, G. (1999) Surface plasmon resonance sensors: review. *Sens. Actuators B*, **54**, 3–15.

[4] Hautakorpi, M., Mattinen, M., and Ludvigsen, H. (2008) Surface-plasmon-resonance sensor based on three-hole microstructured optical fiber. *Opt. Express*, **16**, 8427–8432.

[5] Hassani, A. and Skorobogatiy, M. (2006) Design of the microstructured optical fiber-based surface plasmon resonance sensors with enhanced microfluidics. *Opt. Express*, **14**, 11616–11621.

[6] Hassani, A. and Skorobogatiy, M. (2007) Design criteria for microstructured-optical-fiber-based surface-plasmon-resonance sensors. *J. Opt. Soc. Am. B*, **24**, 1423–1429.

[7] Fu, X., Lu, Y., Huang, X. *et al.* (2011) Surface plasmon resonance sensor based on photonic crystal fiber filled with silver nanowires. *Opt. Appl.*, **41**, 941–951.

[8] P. Yang, Jing, H., and Qisheng, L. (2011) *Simulation of a surface plasmon resonance based on photonic crystal fiber temperature sensor.* Cross Strait Quad-Regional Radio Science and Wireless Technology Conference (CSQRWC), July 26–30, 2011, Heilongjiang, Harbin, China, pp. 274–277.

[9] Qian, W., Zhao, C.-L., He, S. *et al.* (2011) High-sensitivity temperature sensor based on an alcohol-filled photonic crystal fiber loop mirror. *Opt. Lett.*, **36**, 1548–1550.

[10] Qian, W., Chun-liu, Z., Chi Chiu, C. *et al.* (2012) Temperature sensing based on ethanol-filled photonic crystal fiber modal interferometer. *Sens. J., IEEE*, **12**, 2593–2597.

[11] Li, J., Wu, ST., Brugioni, S. *et al.* (2005) Infrared refractive indices of liquid crystals. *J. Appl. Phys.*, **97** (7), 073501–073501-5.

[12] Hameed, M.F.O., Azzam, S.I.H., and Obayya, S.S.A. (2015) Surface plasmon resonance liquid crystal photonic crystal fiber temperature sensor, in *Optical wave and waveguide theory and numerical modelling workshop (OWTNM 2015)*, City University, London.

[13] Obayya, S.S.A., Haxha, S., Rahman, B.M.A., and Grattan, K.T.V. (August 2003) Optimization of optical properties of a deeply-etched semiconductor optical modulator. *J. Lightw. Technol.*, **21** (8), 1813–1819.

[14] Obayya, S.S.A., Rahman, B.M.A., and Grattan, K.T.V. (2005) Accurate finite element modal solution of photonic crystal fibres. *Optoelectron. IEE Proc.*, **152**, 241–246.

[15] Bing, P.B., Li, Z.Y., Yao, J.Q. *et al.* (2012) A photonic crystal fiber based on surface plasmon resonance temperature sensor with liquid core. *Modern Phys. Lett. B*, **26**, 12500821–12500829.

[16] Hameed, M.F.O., Heikal, A.M., Younis, B.M. *et al.* (2015) Ultra-high tunable liquid crystal-plasmonic photonic crystal fiber polarization filter. *Opt. Express*, **23** (6), 7007.

[17] Hameed, M.F.O. and Obayya, S.S.A. (2012) Modal analysis of a novel soft glass photonic crystal fiber with liquid crystal core. *Lightw. Technol. J.*, **30**, 96–102.

[18] Haakestad, M.W., Alkeskjold, T.T., Nielsen, M. *et al.* (2005) Electrically tunable photonic bandgap guidance in a liquid-crystal-filled photonic crystal fiber. *IEEE Photon. Technol. Lett.*, **17** (4), 819–821.

[19] Wei, L., Alkeskjold, T.T., Bjarklev, A. (2010) Tunable and rotatable polarization controller using photonic crystal fiber filled with liquid crystal. *Appl. Phys. Lett.*, **96** (24), 241104.

[20] Akowuah, E.K., Gorman, T., Ademgil, H. *et al.* (2012) Numerical analysis of a photonic crystal fiber for biosensing applications. *Quantum Electron. IEEE J.*, **48**, 1403–1410.

[21] Hassani, A. and Skorobogatiy, M. (2009) Photonic crystal fiber-based plasmonic sensors for the detection of biolayer thickness. *J. Opt. Soc. Am. B*, **26**, 1550–1557.

[22] Obayya, S.S.A. (ed) (2011) *Computational Photonics*, John Wiley & Sons, Hoboken.

[23] Nagasaki, A., Saitoh, K., Koshiba, M. *et al.* (2011) Polarization characteristics of photonic crystal fibers selectively filled with metal wires into cladding air holes. *Opt. Express*, **19**, 3799–3808.

[24] Xue, J., Li, S., Xiao, Y. *et al.* (2013) Polarization filter characters of the gold-coated and the liquid filled photonic crystal fiber based on surface plasmon resonance. *Opt. Express*, **21**, 13733–13740.

[25] Das, S., Dutta, A.J., Patwary, N., Alam, M.S. *et al.* (2013) Characteristic analysis of polarization and dispersion properties of PANDA fiber using finite element methods. *AUST J. Sci. Technol.*, **3**, 3–8.

[26] Kalnins, C., Ebendorff-Heidepriem, H., Spooner, N., and Monro, T. (2012) Radiation dosimetry using optically stimulated luminescence in fluoride phosphate optical fibres. *Opt. Mater. Express*, **2**, 62–70.

[27] Heikal, A.M., Hameed, M.F.O., Obayya, S.S.A. (2013) Improved trenched channel plasmonic waveguide. *J. Lightw. Technol.*, **31**, 2184–2191.

[28] Lei, W., Alkeskjold, T.T., and Bjarklev, A. (2009) Compact design of an electrically tunable and rotatable polarizer based on a liquid crystal photonic bandgap fiber. *Photon. Technol. Lett. IEEE*, **21**, 1633–1635.

[29] Alkeskjold, T., Lægsgaard, J., Bjarklev, A. *et al.* (2004) All-optical modulation in dye-doped nematic liquid crystal photonic bandgap fibers. *Opt. Express*, **12**, 5857–5871.

[30] Hu, C. and Whinnery, J.R. (1974) Losses of a nematic liquid-crystal optical waveguide. *J. Opt. Soc. Am.*, **64**, 1424–1432.

[31] Green, M. and Madden, S.J. (1989) Low loss nematic liquid crystal cored fiber waveguides. *Appl. Opt.*, **28**, 5202–5203.

12

Image Encryption Based on Photonic Liquid Crystal Layers

12.1 Introduction to Optical Image Encryption systems

Encryption is one of the most important and most affordable defenses available to protect our information and, most notably, in securing our online data transmission from attack. Images are widely used in several processes. In recent decades, image encryption techniques have shown great potential in the field of protecting image data from unauthorized access. There are two fundamental ways to encrypt images: symmetric and asymmetric. Symmetric encryption uses an identical key to encrypt and decrypt an image, while in asymmetric encryption, the encryption and decryption keys differ. Symmetric encryption is much faster and less complicated but not as secure as asymmetric encryption; however, because of its speed, it is commonly used to efficiently encrypt large amounts of data [1–3]. Optical security system (OSS) can perform highly accurate encryption and decryption in almost real-time applications. Moreover, OSS provides many degrees of freedom with which the optical beam may be encoded, such as amplitude, phase, wavelength, and polarization. As a result, an optical system employs different schemes, such as amplitude-based encryption, phase-based encryption, or polarization-based encryption. Image encryption schemes based on phase encoding are long lasting and are more secure and flexible than amplitude-based encryption [4–8].

The most widespread approach in optical encryption systems, proposed by Refregier and Javidi, is double random phase encoding (DRPE) and is described in Ref. [9]. As proposed in Ref. [9], although the system performance is satisfactory and allows one to obtain encrypted images that are characterized by high cryptography resistance, the use of lenses makes it less appealing for practical purposes as it needs precise adjustment. Also, DRPE does not offer fully functionally integrated systems since it needs a propagation distance equal to the focal length of lens. Moreover, several drawbacks of such systems such as complexity of the optical key diagram and expensive have been reported recently [10, 11].

Currently, there are many available developed approaches for optical encryption systems [9–13]. The concept of modes on DRPE-type encryption systems has been introduced to analyze the encryption system in the context of known attacks [12]. Cryptographic block ciphers

Computational Liquid Crystal Photonics: Fundamentals, Modelling and Applications, First Edition.
Salah Obayya, Mohamed Farhat O. Hameed and Nihal F.F. Areed.
© 2016 John Wiley & Sons, Ltd. Published 2016 by John Wiley & Sons, Ltd.

partition messages into data blocks before transmission. These blocks are then processed one at a time. The best way to do this is to use the standard modes of operation together with the basic cryptographic algorithm. These modes of operation can be used to pad in a more secure way, control error propagation, and transform a block cipher into an arbitrary length stream cipher. Although the algorithms reported in Ref. [12] describe several modes of operation with increased sophistication that allow the sender some level of defense against known attacks upon DRPE. However, the additional security that arises from employing these algorithms in physical optical systems was not considered. Further, radically different considerations for optical encryption based on the concepts of computational ghost imaging have been proposed [13]. Ghost imaging is an intriguing optical technique where the imaging information is obtained through photon coincidence detection. The idea is based on sharing a secret key, consisting of a vector of N components between the sender and the recipient. In addition, a spatially coherent monochromatic laser beam passes through a spatial light modulator, which introduces an arbitrary phase-only mask. The transmitted light is collected by a single-pixel detector. This operation is repeated N times for N different phase profiles, each of them corresponding to one secret key component. The encrypted version of the object image is not a complex-valued matrix but simply an intensity vector, which noticeably reduces the number of transmitted bits. Moreover, a three-dimensional (3D)-space-based approach is used as a new method for optical encryption in various encryption architectures [14]. The fundamental feature of this proposed method is that each pixel of the plaintext is axially translated and considered as one particle in the proposed space-based optical image encryption, and the diffraction of all particles forms an object wave in the phase-shifting digital holography. In Ref. [15], one of the latest developments in encryption technology is introduced. It involves a new method using a structured illumination-based diffractive imaging with a laterally translated phase grating for optical double-image cryptography. Additionally, the integration of the photon-counting imaging technique with optical encryption has been recently proposed to obtain a photon-limited version of the encrypted distribution [16]. Although the proposed systems reported in Refs. [12–16] offer high robustness against ciphertext contaminations, the offered systems are not suitable for real-time encryption and also lack tunability and fabrication simplicity.

Photonic crystals (PhCs) or photonic bandgap (PBG) structures are attractive optical materials for controlling and manipulating the flow of light. Planar PhC provides a promising platform for the next-generation integrated optical circuits that allow high performance and also offer the possibility of integration with electronic circuits [17]. Therefore, a problem of integration raised by the conventional DRPE encryption system can be easily solved by using PBG structures, since all components will be designed around the same PhC platform. 1D PhC is already in widespread use in applications of image encryption and hiding systems [18, 19]. He *et al.* [19] demonstrated a method for multiple-image hiding on the basis of interference-based encryption architecture and grating modulation. In addition, a new idea for data-image encryption and decryption based on a 3D PhC structure has been reported by Areed *et al.* [20]. In Ref. [20], the phase of the optical image can be encoded by adding a constant phase shift 180° within the frequency band of the image; while to decode the image, a complementary phase shift 180° will be added. Areed *et al.* suggested a PhC structure with reflection properties of unity magnitude and 180° phase shift. This can be achieved by forming a cubic lattice array of air spherical holes in a cube of dielectric with permittivity $\varepsilon_r = 10.8$. Although the suggested design in Ref. [20] is effectively applied to realize for the first time, real-time image encryption with flexible integration, the offered design lacks tunability for varying the encoding phase, parallel processing for simultaneous encryption, and also requires a complex fabrication process. A detailed description of the proposed system will follow in Section 12.2. Moreover, Areed *et al.* have developed the structure in Ref. [20] to realize a simultaneous multiple image encryption/decryption system by integrating the 2D PhC and nematic liquid crystal (NLC) layers [21]. A detailed description of the proposed system in Ref. [21] will follow in Section 12.3.

12.2   Symmetric Encryption Using PhC Structures

In this section, the idea for data-image encryption and decryption that relies on the use of 3D PBG structures [20] will be discussed in detail. These structures are periodic and are used to control many features of electromagnetic radiation in certain bands of frequencies. The emerging technology of 1D, 2D, and 3D PBG structures can be suitably exploited to design and make optical devices, such as waveguides, splitters, resonant cavities, and filters [17]. The proposed design in Ref. [20] adopts a carefully tailored PBG block that exhibits high reflectivity and constant phase properties within our frequency range of interest. The design of the proposed 3D PBG encryptor depends on studying the relation between the reflection properties and the geometrical parameters of the 1D PBG structure. The use of the 3D PBG enhances the security of the encoding system by replacing the complicated encryption–decryption designs by only one stage of a 3D hardware key. Additionally, the proposed optical 3D encryptor offers 86 ps response time that facilitates the correct sampling of the signal and signal recovery for online encryption–decryption systems. In Ref. [20], to demonstrate the excellent performance of the optical symmetric encryption system, the mean square error between the decrypted and original image has been calculated. Although used at microwave frequencies, the offered design can be easily extended to the optical frequency range via appropriate scaling of the system dimensions. Due to its significantly better performance, the designed PBG encryption/decryption approach renders itself as a highly competitive approach in comparison with recently existing encryption techniques based on diffraction gratings [22].

A 3D simulator based on the finite integration time domain (FITD) algorithm was used to simulate the PBG structures. The FITD is a consistent formulation for the discrete representation of the integral form of Maxwell's equations on spatial grids. The FITD was proposed by Weiland [23, 24] in 1977. The perfect boundary approximation (PBA) technique that relies on nonuniform meshing applied in conjunction with FITD maintains all the advantages of the structured Cartesian grids, while allowing the accurate meshing around the curved boundaries: a crucial point for modeling PBG structure [25].

This section is organized as follows. Following this introduction, some analysis and temporal response of the 1D PBG structure is given in Section 12.2.1. In Section 12.2.2, the generalized 3D PBG encryption/decryption structure is explained. Section 12.2.3 presents the simulation results showing the performance of the proposed structure [20] in the encryption and decryption of an image.

12.2.1   Design Concept

First, we consider six periods of Bragg grating structure made of RT/duroid 6010LM ($\varepsilon_r = 10.8$) excited by an electric mode located at 39 mm from its boundary. According to the simple equation relating the Bragg reflected wavelength λ, the effective refractive index n_{eff}, and the

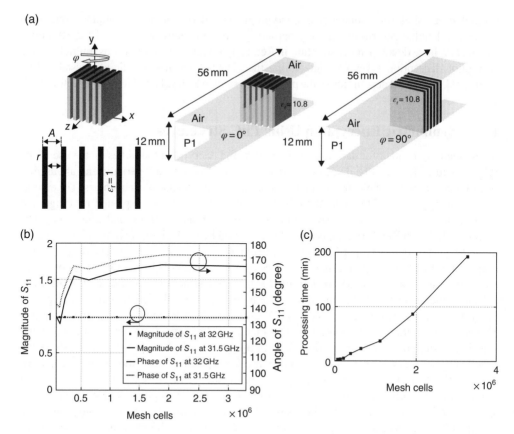

Figure 12.1 Six period of Bragg grating: (a) details of the structure, (b) reflection coefficients versus the number of mesh cells, and (c) computational time versus the number of mesh cells

grating period A ($\lambda = 2n_{eff}A$) [26], the reflection wavelength range will be around 34 GHz for a refractive index 3.286, which is essentially in the same frequency range considered in Ref. [20].

The FITD is used to calculate the transmitted and reflected power spectra of the considered structure whose schematic diagram is shown in Figure 12.1a. Because of the symmetrical nature of the system, only one-quarter of the computational domain size ($x \times y \times z$) (6 mm × 6 mm × 56 mm) structure is analyzed. The time variations of the fields are recorded at the reference ports P1 and P2, respectively. Shown in Figure 12.1b are the results of variation of S_{11} with different mesh densities, and the results clearly demonstrate the numerical convergence of the adopted FITD. It can be noted from the plot that, to keep reasonable accuracy, this structure has been discretized with a mesh cell size of $\lambda/30$ or less. On a personal computer; (Pentium IV, 3.2 GHz, 2 GB RAM), the whole simulation for the perfect meshing period took around 3 h as shown in Figure 12.1c.

The influence of the r/A ratio of the grating has been investigated considering the geometry with $\varphi = 0°$ where r is the thickness of the RT/duroid layer in one period. Figure 12.2 shows the calculated magnitude and phase of the reflection coefficient of the grating with r/A 0.1–0.9 for the frequency range 30–34 GHz. It may be observed from Figure 12.2a that high reflected

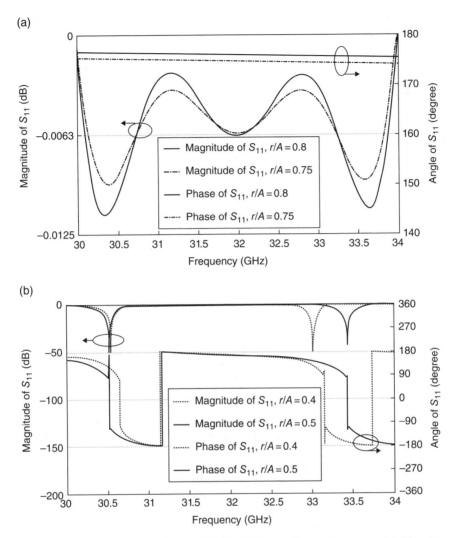

Figure 12.2 Simulated S_{11} parameter for different r/A ratios: (a) $r/A > 0.7$ and (b) $r/A \leq 0.5$

power and a constant S_{11} phase are obtained with the grating characterized by large r/A (0.75 or more), while Figure 12.2b shows that a variable S_{11} phase is obtained with the grating characterized by small r/A (<0.7). The bandwidth of the constant S_{11} phase is calculated and plotted versus r/A in Figure 12.3a. It can be seen that the angles of S_{11} at different frequencies tend to be the same at $r/A > 0.7$. Symmetrical grating configurations at different rotation angle φ have been examined. Figure 12.3b shows the calculated decibel magnitude and phase curves of the reflection coefficients for $\varphi = 0°$ and $\varphi = 90°$. It can be seen that the phase values of S_{11} increase with increasing the rotation angle φ.

It can be seen from Figure 12.3c how the phase of S_{11} characteristic of the grating ($\varphi = 90°$) changes as a function of the lattice constant A. It is noticed that decreasing A causes a significant increase in the S_{11} phase values.

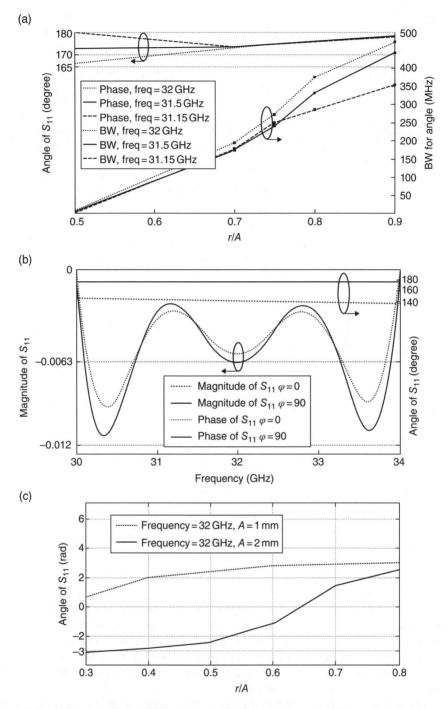

Figure 12.3 (a) Bandwidth of constant phase, simulated S_{11}-parameter for different values of (b) rotation angle φ, and (c) lattice constant A

12.2.2 Encryptor/Decryptor Design

The already analyzed 1D PBG is extended to the 3D PBG case in order to make an image encryption scheme. Figure 12.4 shows the optical encryption structure that embodies a cubic lattice array of spherical air holes of 1 mm lattice spacing using a dielectric substrate with dielectric constant 10.8.

As observed from the analysis of the 1D structure, the phase variation is highly sensitive to the orientation of the Bragg-grating plates. However, by employing a symmetric 3D PBG structure, the waves impinging on the structure will always face the same scatters, that is, the phase relationship will remain unchanged which is the main advantage of the 3D PBG over the 1D component. The PBG of different configurations with different spherical hole diameters are tested by calculating the reflection coefficient (S_{11}) as shown in Figure 12.5. In this investigations, the lattice constant is taken as $A=1$ mm. The figure shows that, the hole diameter $r=0.7$ mm, results in a stop band (S_{11} magnitude near 0 dB and variable S_{11} phase) over 32.532–32.535 GHz, whereas the hole diameter $r=0.9$ mm, results in a stop band (S_{11} magnitude near 0 dB and constant S_{11} phase 127°) over 32.51–32.55 GHz, respectively. The curves prove that a large bandwidth of the complete bandgap with constant phase is obtained by a large r/A ratio, as previously studied.

Next, the lattice spacing tuning is applied to increase the S_{11} phase from 127° to 180° as shown in Figure 12.6a. Figure 12.6b shows the calculated magnitude and the phase of the scattering parameter S_{11} which are nearly 0 dB and 180° over 31.1–31.19 GHz, at a lattice spacing $A=0.8$ mm and hole diameter $r=0.72$ mm. Figure 12.7 shows the steady-state real part of the y-polarized electric field along the propagation direction z at the central frequency 31.095 GHz at $r=0.72$ mm $A=0.8$ mm. Computation time for the displayed example is about 2 h. It is evident from Figure 12.7 that the reflected wave along the structure is highly confined and also reaches the steady-state sinusoidal variations at frequency of 31.095 GHz, which lies within the operating bandwidth. This reflected wave confinement clearly agrees with the behavior of S_{11} shown earlier in Figure 12.6b.

As discussed previously, the optimum design of the proposed encryptor [20] is at $A=0.8$ mm where a phase of 180° is obtained. However, in an attempt to work out beyond this value of A, the wave will be lost completely due to the simple fact that it is impossible to catch any reflected waves out of the PBG range. The 3D PBG technology has become much more mature in the recent years, spanning a number of application areas, such as telecommunication, sensing, filtering, and now, for the first time to the best of our knowledge, in encryption, as suggested in this chapter. The authors of Ref. [20] believe that the fabrication of 3D PBG is not difficult and has already been implemented [27].

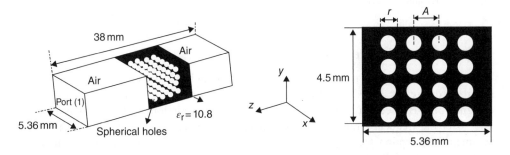

Figure 12.4 3D photonic bandgap encryptor: (a) details of the 3D structure and (b) transverse 2D plane

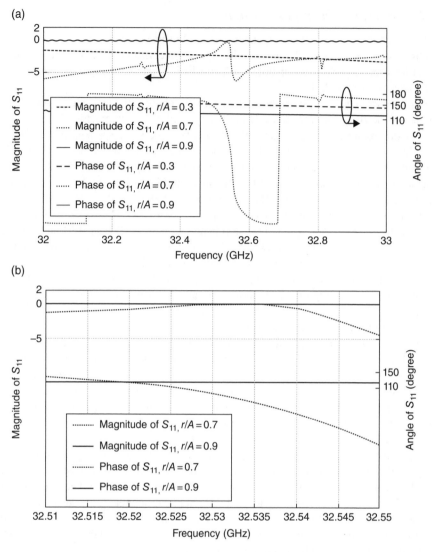

Figure 12.5 Simulated S_{11}-parameter for different r/A ratios, $A = 0.8$ mm versus the frequency bands: (a) 32–33 GHz and (b) 32.15–32.55 GHz

12.2.3   Simulation Results

The main idea behind the proposed method to encrypt digital images is to create an easy and secure hardware key. Areed *et al.* propose a new encryption technique that uses the previously designed 3D PBG section as a band-stop filter (zero transmission) with a constant phase of 180° over the frequency range of the modulated image. The architecture of the proposed symmetric encryption technique is depicted in Figure 12.8. The encryption technique is realized using the following steps:

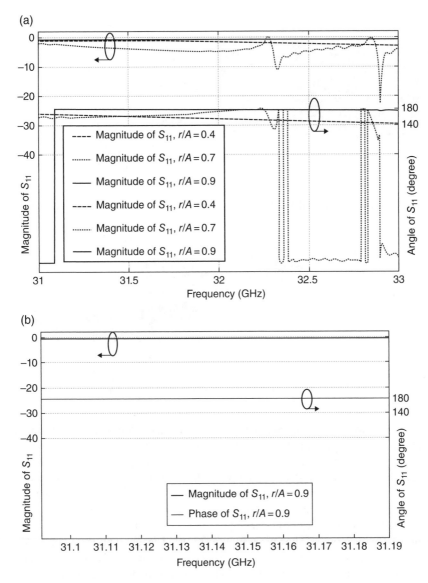

Figure 12.6 Simulated S_{11}-parameter for different r/A ratios, $A = 0.8\,\text{mm}$ versus the frequency bands: (a) 31–33 GHz and (b) 31.1–31.2 GHz

Step1: Modulate the source image to a central carrier frequency 31.14 GHz.

Step2: Apply the modulated image to the designed 3D PBG block with reflection results given in Figure 12.6b through an isolator. The isolator device is placed to prevent the PBG reflected wave from passing to the modulator block. At this point, the signal will be totally reflected with its phase modified by 180° over the frequency range.

Step3: Apply the reflected signal to the demodulator block to obtain the encrypted image.

(a)

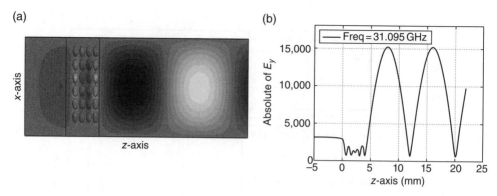

(b)

Figure 12.7   Steady-state field profile at $f = 31.095\,\text{GHz}$ versus: (a) the xz-plane and (b) the z-axis

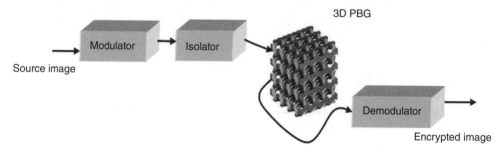

Figure 12.8   Architecture of symmetric key encryption

It is worth mentioning that the modulator and demodulator blocks are merely simple codes programmed on a computer unit to deal with the digital image.

To retrieve the decrypted image, the same block diagram can be used where the encrypted image will be the input to the system. The output of the decryption phase will be the source image with its phase modified by 360°, that is approximately zero. Thanks to the adopted symmetric encryption approach, the overall encryption/decryption system is easy to implement.

To evaluate the optical encryption system, this method is tested on a gray-level cameraman image of [256×256] pixels shown in Figure 12.9a. The encryption result is shown in Figure 12.9b. The accuracy in retrieving the images depends extremely on the fabrication process and can be estimated by calculating the values of the root mean square (rms) of error and rms signal-to-noise ratio (SNR)$_{\text{rms}}$. These criteria quantify the error between the source image and the decrypted image as can be described by the following relations [28]:

$$\text{rms} = \sqrt{\frac{1}{N^2}\sum_{i=1}^{N}\sum_{j=1}^{N}\left[g(i,j) - f(i,j)\right]^2}. \tag{12.1}$$

$$\text{SNR}_{\text{rms}} = \sqrt{\frac{\sum_{i=1}^{N}\sum_{j=1}^{N}g^2(i,j)}{\sum_{i=1}^{N}\sum_{j=1}^{N}\left[g(i,j) - f(i,j)\right]^2}}. \tag{12.2}$$

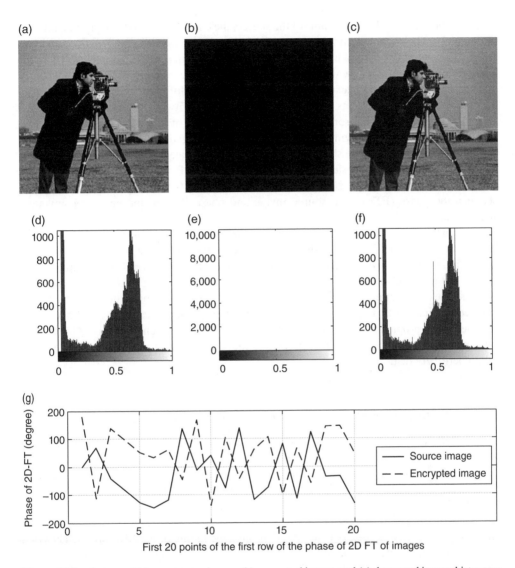

Figure 12.9 Source of (a) cameraman image, (b) encrypted image, and (c) decrypted image; histogram of the (d) source image, (e) encrypted image, and (f) decrypted image; and (g) phase samples of the 2D FT of the source and encrypted images

Here, $g(i,j)$ is the pixel intensity of the original image and $f(i,j)$ is the pixel intensity of the decrypted image. The row and column numbers of these two images are defined by N. Again to determine the quality of the decrypted image, the difference between the original image and the decrypted image is computed.

Using the data of the original image of Figure 12.9a and the data of the retrieved image shown in Figure 12.9c, the values of the rms and SNR_{rms} are equal to 0.0759 and 6.426, respectively.

Figure 12.9d–f shows the histograms of the source, the encrypted, and the decrypted images, respectively. The obtained simulation results of the encryptor design of SNR_{rms} and rms clearly agree with the plotted histograms of the source and the decrypted images where the histograms are approximately the same. Moreover, it may be noted from Figure 12.9d and e that the histogram of the encrypted image is completely different from the histogram of the source image. This reflects the efficiency and good quality of our suggested encryption system. It may be observed from these histograms that most of the encrypted image tends to lie within the black color, giving rise to its pixels taking negative values. This negative sign is due to the added 180° phase shift by the 3D PBG block as clearly illustrated in Figure 12.9g which shows the computed samples of phases of the 2D Fourier transform (FT) for the source and the encrypted images. It may be noted from the figure that the phase difference between the phase (FT) of the source image and phase (FT) of the encrypted image is approximately 180°.

To evaluate the system in the case of attacks, the noise added by encryption to the source image has been quantified as the difference between the encrypted image power and the source image power. By normalizing this encryption noise power to the source image power, we have achieved SNR as good as −3 dB; therefore, the system is reasonably immune to external attack. The computational time of the encryption or decryption process has been estimated to be approximately 86 ps which is the time taken by the wave to complete one propagation trip (forward and backward) through the PBG block. This extremely low computational time reflects the fact that the encryption model in Ref. [20] is competitively fast.

12.3   Multiple Encryption System Using Photonic LC Layers

In this section, we will discuss in detail the integration of 2D PhC and NLC layers that have been suggested in Ref. [21] to realize simultaneously a multiple image encryption/ decryption system. The parameters of the 2D PhC structure have been carefully selected in order to realize the PBG effect that can be used to achieve complete isolation between two NLC layers. The proposed idea in Ref. [20] for data-image encryption and decryption based on phase encoding has been adopted in this work. However, in our design, the required phase shifts can be easily achieved through the tuning of the NLC layers. The NLC has received much interest due to its external-field-dependent optical anisotropy and high birefringence. This property of the NLC allows a high degree of freedom in easily obtaining either phase shift $\psi°$ for encryption or the complementary phase shift $-\psi°$ for decryption. The use of the proposed scheme enhances the security of the encoding system as the pixels of two images can be simultaneously encrypted/decrypted by only one PBG block placed between two NLC layers. Moreover, the system tunabliity can be easily achieved through a random jump from phase $\psi_1°$ to $\psi_2°$ over the same system. In addition, the suggested structure in Ref. [21] with faster and efficient performance is more easily fabricated than the proposed 3D PhC in Ref. [20]. The accuracy of the proposed encryption system in Ref. [21] has been quantified by calculating the rms and SNR_{rms} between the decrypted and original images.

The section is organized as follows. Following this introduction, a detailed description of the proposed system will follow in Section 12.3.1. Section 13.3.2 presents the simulation results of the proposed encryption/decryption system of Ref. [21]. The performance of the proposed system in image encryption and decryption is also tested and quantified.

12.3.1 Proposed Encryption System

Based on the idea of the phase-shifted data-image explained in the previous sections, a novel architecture that consists of a 2D PBG layer placed between two NLC layers was proposed for the first time in Ref. [21]. The system block diagram shown in Figure 12.10 clearly has the advantage of offering the efficient use of the hardware to encrypt two images at the same time. The design of the 2D PBG structure is optimized to obtain high reflectivity and also to isolate the image data flow from the two image sources 1 and 2.

12.3.1.1 PBG Structure

The proposed encryption system shown in Figure 12.10 is built with the aid of the 2D PhC structure suggested in Ref. [29]. The 2D PhC structure is composed of a 2D square lattice array of high-index dielectric rods, each rod has a refractive index of 3.4 and radius of $0.18a$, where a is the lattice constant. The geometrical parameters of the 2D PhC structure have been carefully chosen in order to realize the PBG effect and to isolate the image data flow from the two image sources 1 and 2. The functions of the proposed 2D PhC blocks in the regions near the ports are to create two separate paths for the incident and reflected laser beams. The fabrication of the utilized 2D PhC is not so difficult, and a similar structure has already been made as reported in Ref. [30].

12.3.1.2 Liquid Crystals

LCs are anisotropic materials where their electrical permittivity is highly dependent on the orientation of their constituent molecules. When applying an external electrical field, the molecules of the LC start to align along the field lines. The magnitude of the orientation depends on the field strength. The electro-optic properties of the LCs can be suitably exploited to design highly tunable photonic devices. In this investigation [21], two NLC layers of type E7 are placed in quartz pans. Two alignment layers, each of 100 nm polyimide are placed on the bottom surface of the quartz walls along the y-axis. The biasing states of the NLC layers can be modified using two transparent electrodes. The relative permittivity tensor ε_r of the E7 material is taken as [31, 32]

$$\varepsilon_r = \begin{pmatrix} n_o^2 \sin^2\varphi + n_e^2 \cos^2\varphi & \left(n_e^2 - n_o^2\right)\cos\varphi\sin\varphi & 0 \\ \left(n_e^2 - n_o^2\right)\cos\varphi\sin\varphi & n_o^2 \cos^2\varphi + n_e^2 \sin^2\varphi & 0 \\ 0 & 0 & n_o^2 \end{pmatrix}, \tag{12.3}$$

where φ is the rotation angle of the director of the NLC and n_o and n_e are the ordinary and extraordinary refractive indices of the E7 material and can be calculated using the Cauchy model [31].

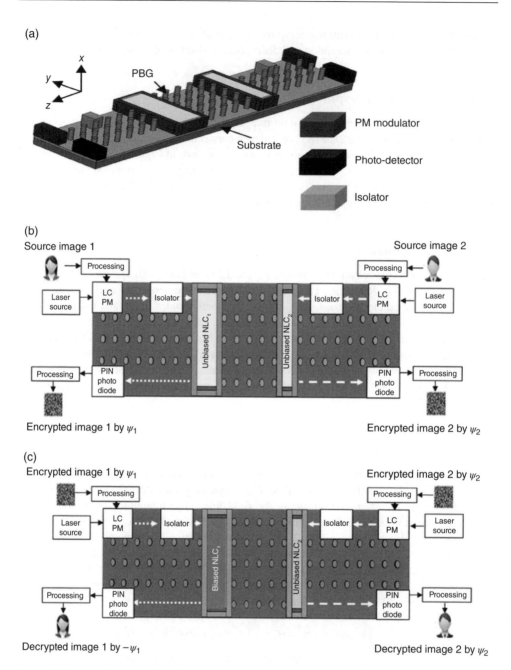

Figure 12.10 Architecture of multiple encryption system: (a) 3D view, (b) stage of encryption, and (c) stage of decryption

12.3.1.3   Phase Modulator/Photodetector

A phase modulator (PM) is an optical modulator which can be used to control the optical phase of a laser beam by an input microwave signal (processed analog image). Frequently used types of PM within the area of integrated optics are electro-optic modulators based on Pockels cells and LC modulators [33]. By exposing a crystal modulator, such as lithium niobate, to an electric field, the phase of the laser light exiting the crystal can be controlled by changing the electric field in the crystal. The reflected laser light from the PBG block illuminates the PIN photodiode to produce an electrical signal that is further processed to recover the analog image. There are many ultrafast photodetectors that can be easily integrated and designed upon the considered PhC platform [34].

12.3.1.4   System Operation

The following step-by-step description may help explain the proposed technique in Ref. [21] to encrypt two images simultaneously:

- Step1: Modulate the phases of the two laser beams that each has a selected wavelength from the central zone of the PBG range by the pixel intensities of the two source images.
- Step2: Apply the modulated image 1 to the proposed encryptor from the left side. The modulated image 1 will propagate through the first unbiased NLC_1 layer and be totally reflected due to the PBG block with its phase modified by $-90°$. Simultaneously, apply the modulated image 2 to the proposed encryptor from the right side. The modulated image 2 will propagate through the second unbiased NLC_2 backed by the PBG block and be totally reflected with its phase modified by $-180°$ over the image 2 bandwidth. The required phase shifts can be easily achieved through optimizing the thickness of the NLC layers. The isolator is used to prevent the reflected encoded laser beam being reapplied to the modulator block. The isolator can be made compact either by an ultra-fast LC tunable switch, as reported in Ref. [35] by changing the structure of the PhC as reported in Ref. [36].
- Step3: The reflected laser beams from the PBG block illuminate the PIN photodiodes to produce electrical signals that are further processed to recover the analog encrypted images.

To regain the decrypted images, the same encryptor design can be used after modifying only the biasing state of the first NLC_1 layer to obtain the complementary phase shift $90°$ instead of $-90°$. However, the unbiased NLC_2 should be maintained to obtain phase shift of about $-180°$. At this point, the recovered images from the decryption stage will be the source images with their phases modified by approximately zero. Due to the tunability of the NLC layers, the overall encryption/decryption system is remarkable to implement.

12.3.2   Simulation Results

First, the 3D finite-difference time domain (FDTD) [37] numerical method was applied to investigate the PBG of a 2D PhC structure shown in Figure 12.11a. This figure demonstrates a 2D array of silicon dielectric rods built on a silicon substrate that is isolated by a 200 nm layer of low index Al_xO_y. Each rod in the 2D PhC layer has a radius of 0.1044 μm, height of 1 μm and is placed in a square lattice with a period of 0.58 μm. Thanks to the symmetrical nature of the

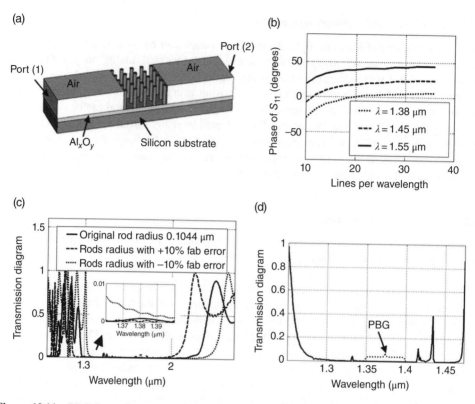

Figure 12.11 PBG for an array of 5×5 silicon rods: (a) geometry details, (b) argument of S_{11} versus the lines per wavelength, transmission power: (c) without substrate and (d) with substrate

device, only one-half of the computational domain $(x \times y \times z)$ $(0.5 \,\mu m \times 3.86 \,\mu m \times 10.528 \,\mu m)$ of the structure is analyzed. Figure 12.11b shows the variation of the S_{11} argument with different mesh sizes and the results clearly demonstrate that to attain acceptable accuracy the structure should be discretized with a mesh cell size of $\lambda/30$ or less. Figure 12.11c demonstrates the transmission characteristics versus the wavelength across the 2D PhC layer without substrate. It shows that a PBG from 1.3 to $2 \,\mu m$ is created. Moreover, $\pm 10\%$ changes in the rod radius result in only a 1% drop in the overall reflectivity of the PBG block relative to the original. It is worth mentioning that the main role of the PBG block is to act as an almost ideal reflector to isolate the two image sources. Figure 12.11d shows the transmission characteristics across the 2D PhC layer with isolated substrate. The figure shows that the PBG effect has been realized within the range 1.35–1.4 μm.

Next, the previously analyzed 2D PhC layer is backed by an NLC layer of type E7, as shown in Figure 12.12a. In the simulation, the NLC layer has length 2.52 μm, width d, and height 1 μm. The quartz pan is used to create a cavity that is filled with the NLC. In our calculations, the effect of the quartz pan has been taken into account as evident from Figure 12.12b. The thickness of the quartz layers has been chosen so that the overall performance of the system is unchanged. Two alignment layers, each of 100 nm polyimide, are placed on the bottom surface of the quartz walls along the y-axis.

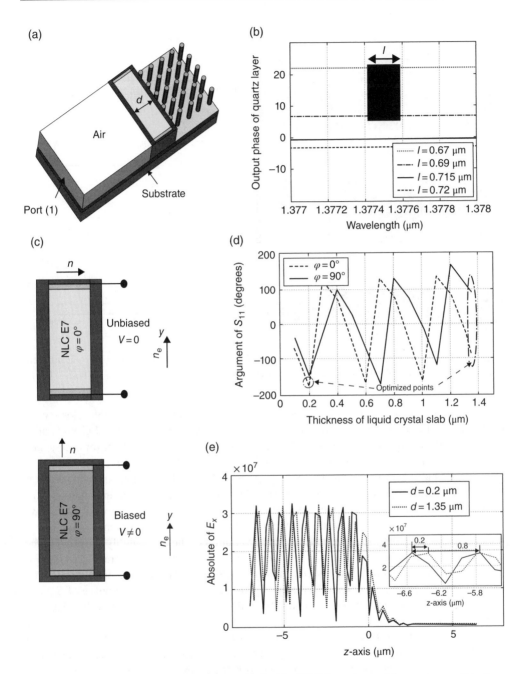

Figure 12.12 (a) Array of 5×5 silicon rods backed by NLC layer inserted in quartz pan, (b) phase effect of quartz layer, (c) top views for LC molecular orientation for unbiased and biased states, (d) argument of S_{11} versus the NLC thickness d, and (e) steady-state field profile at $\lambda = 1.3772\,\mu m$

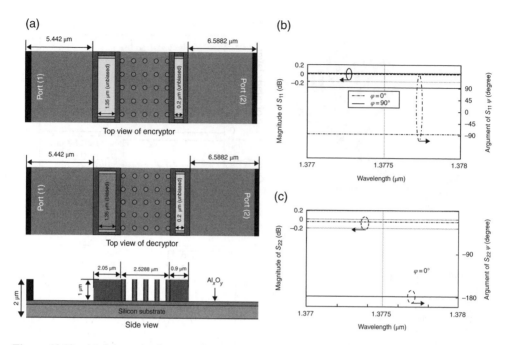

Figure 12.13 (a) Schematic diagram of the proposed encryptor/decryptor system, (b) S_{11} at different rotation angles $0°$ and $90°$, and (c) S_{22} at rotation angle $\varphi = 0°$

Figure 12.12c demonstrates two top views for the orientations of the LC molecules for the unbiased and biased states. In the case of the unbiased state ($V_{LC} = 0$), the LC molecules gradually stand along the x-axis with rotation angle $\varphi = 0°$. Therefore, the light for both y- and z-polarizations is influenced by the ordinary index n_o of the LC cell. However, in the case of the saturated biasing state, the voltage applied across the biasing layers along the y-axis rotates the LC molecules perpendicular to the cell wall along the y-axis, with rotation angle $\varphi = 90°$. Therefore, the y-polarized light depends on the extraordinary index n_e. Figure 12.12d shows the variation of the S_{11} argument with the thickness of the NLC layer at different biasing states. It can be observed from the figure that the phase variation of the S_{11} is highly sensitive to the thickness of the NLC in a periodic manner. Moreover, it is evident from Figure 12.12d that for the unbiased NLC with thickness $0.2\,\mu m$, the incident wave will be totally reflected with its phase modified by $\psi = -180°$. Again, Figure 12.12d shows that the phase of S_{11} will be ($-90°$ in the unbiased state $\varphi = 0°$) and ($90°$ in the biased state $\varphi = 90°$) for NLC layer of thickness $1.35\,\mu m$.

Figure 12.12e demonstrates the steady-state distributions of the x-polarized electric field along the propagation direction z at wavelength of $1.3772\,\mu m$ for the selected NLC thicknesses 0.2 and $1.35\,\mu m$, respectively. It can be observed from the figure that the reflected waves along the structure are highly bent with steady-state sinusoidal variations. Moreover, the reflected wave at $d = 1.35\,\mu m$ overrides the reflected wave at $d = 0.2\,\mu m$ by about $\lambda/4$, which clearly fits with the behavior of the argument of S_{11} shown in Figure 12.12d. Figure 12.13a shows the schematic diagrams for the proposed design of an encryptor/decryptor system. The structure is composed of a PBG block which is backed from the left by the first NLC_1 layer of thickness

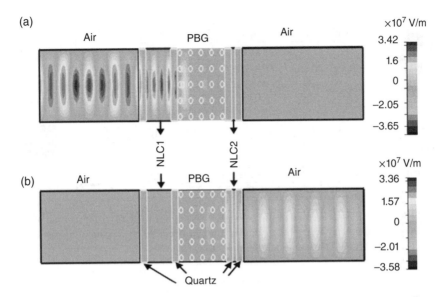

Figure 12.14 2D steady-state field profile at $\lambda = 1.3772\,\mu m$ when inserting the signal at (a) port (1) and (b) port (2)

1.35 μm, while the second NLC_2 layer of thickness 0.2 μm is placed on the right side of the PBG block. Each of the two NLC layers is placed in a quartz pan. Figure 12.13b, c shows the scattering parameters S_{11} and S_{22} as a function of wavelength across the proposed encryption system shown in Figure 12.13a. It is evident from the figures that using both NLC layers with zero biasing state ($\varphi = 0°$) results in a stop band (S_{11} magnitude of −0.02831 dB with constant S_{11} phase of −89.9°) and (S_{22} magnitude of −0.0529 dB with constant S_{22} phase of −179°) over the wavelength range 1.377–1.378 μm. Whereas using a biased NLC_1 layer ($\varphi = 90°$), results in a stop band (S_{11} magnitude of −0.01841 dB with constant S_{11} phase of +90.9°) over the wavelength range 1.377–1.378 μm, respectively. The tunability of the NLC_1 can be easily noticed, where changing the rotation angle φ from 0° to 90° varies the phase ψ of the reflected wave from −90° to +90°. It may be observed that the proposed design can be employed to simultaneously encrypt/decrypt two images.

Figure 12.14 illustrates the steady-state distributions of the x-polarized electric field along the yz-plane at the central wavelength 1.3772 μm for the proposed encryptor shown in Figure 12.13a. It is obvious from this figure that the reflected waves along the structure are highly confined with steady-state sinusoidal variations. These reflected waves clearly match with the behaviors of S_{11} and S_{22} shown previously in Figure 12.13b and c, respectively.

In Ref. [21], the proposed encryption system has been tested using two different gray-level images, each of [3×3] pixels. The selection of a relatively small-sized image in Ref. [21] is merely for the purpose of proof of concept. Assuming four different pixels intensities of 0, 0.3, 0.7, and 1 are inputted to the PM shown in Figure 12.10, the emerging laser beam will encounter phases of 0°, +90°, −90°, and180°, respectively. Moreover, this laser beam itself experiences another −90° phase shift or −180° phase shift resulting from reflection from the PBG block, as evident from Figure 12.10. At this point, we can follow the encryption procedure for encrypting the source image 1 as explained in the sequence and shown in Figure 12.15.

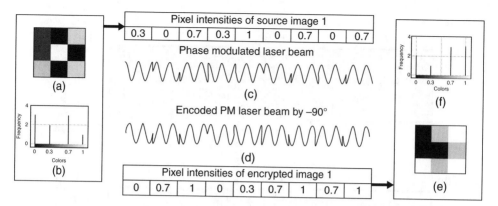

Figure 12.15 Encryption procedure of source image 1: (a) source image 1, (b) histogram of the source image 1, (c) output laser beam from phase modulator, (d) encoded PM laser beam using phase shift –90°, (e) encrypted image 1, and (f) histogram of the encrypted image 1

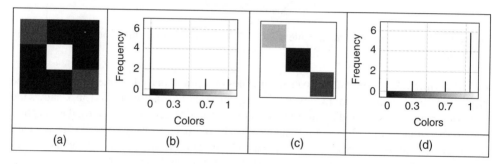

Figure 12.16 (a) Source image 2, (b) histogram of the source image 2, (c) encrypted image 2 using phase shift –180°, and (d) histogram of the encrypted image 2

Shown in Figure 12.15a is the source image 1 where its histogram shows a frequency content of the colors as revealed from Figure 12.15b. From this figure, it is quite evident that the pixels are quite concentrated around the aforementioned colors. Next, Figure 12.15c demonstrates that different pixels have evidently changed laser beam phases, while the –90° phase shift due to reflection from the carefully designed PBG and NLC_1 has been clearly illustrated by the laser beam depicted in Figure 12.15d. Next, the reflected encoded laser beam form the PBG block illuminates the PIN photodiode to produce an electrical signal that is further processed to recover the encrypted image 1 shown in Figure 12.15e. The histogram of the encrypted image shown in Figure 12.15f is apparently shifted from its counterpart of the original image shown in Figure 12.15b. Simultaneously, the source image 2 will be encoded in a very similar way except that the laser beam propagating in the right-hand section of the system experiences a phase shift of –180° due to reflection from the carefully designed PBG and NLC_2. In addition, quite similar to the diagrams of Figure 12.15, the effect of encrypting source image 2 is quite clear from the diagrams shown in Figure 12.16, where the shift of the pixel frequency contents is evident from Figure 12.16b and d. The amounts of the noise added by the proposed encryption system to the source images 1 and 2 have been quantified as the difference between

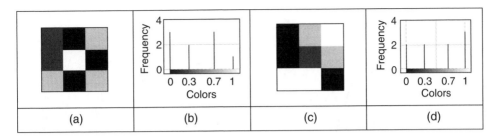

Figure 12.17 (a) Source image 1, (b) histogram of the source image 1, (c) encrypted image 1 using phase shifts –180° and –90°, and (d) histogram of the encrypted image 1

the encrypted image power and the source image power. By normalizing the power of the encoding noise to the power of the source image, we have attained SNR as good as –1.29 and 6.658 dB for the encryption using phase shifts –90° and –180°, respectively. Figure 12.17 shows the encryption of source image 1 where the first seven pixels are encrypted using a phase shift of –90°, while the other two pixels are encrypted using a phase shift of –180°. The correlation coefficients between the source image 1 shown in Figure 12.15a and its encrypted images shown in Figures 12.15e and 12.17c are equal to –0.1475 and –0.1165, respectively.

The estimated values of the correlation coefficients indicate that the encryption using a random jump from phase –90° to –180° over the same system is better than the encryption using a single phase shift. These relatively small-sized images are merely for the purpose of demonstrating the encryption idea. However, with four available phases {0°, –90°, 90°, 180°}, we have the potential of 2^4 different phase combinations to images with a large number of colors.

The average time of the encryption process or decryption process is the measure of the time taken by the wave to finish one propagation travel (forward and backward) through the encryption or decryption stage and is calculated to be approximately 700 fs. The proposed system with extremely low processing time can be suitably exploited for real time optical image encryption systems.

The recovered decrypted images can be obtained using the same encryptor design after modifying only the biasing state of the first NLC_1 layer to obtain the complementary phase shift 90° instead of –90° and maintaining the unbiased NLC_2 as it is to obtain a phase shift of about –180°. The accuracy in retrieving the images depends extremely on the fabrication process and can be estimated by calculating the values of the rms of error and SNR_{rms}. These criteria quantify the error between the source image and the decrypted image as can be described by Eqs. (12.1) and (12.2) [28].

The estimated results of the rms and SNR_{rms} for the two decoded images are rms = 0.0051 and SNR_{rms} = 81 dB for encryption using ψ (–90° and 90°) and rms = 0.0057, SNR_{rms} = 74 dB for encryption using ψ (–180°).

The tolerances with respect to the error of fabrication have been evaluated. During the fabrication process, errors can be made simultaneously on several parameters, especially on the NLC thickness and the rotation angle φ. System rms and SNR_{rms} versus the variation ranges of the NLC thickness and the rotation angle have been evaluated and plotted, as shown in Figure 12.18. As may be observed, the overall system rms has been slightly increased, while the overall system SNR_{rms} has been slightly decreased due to possible fabrication errors.

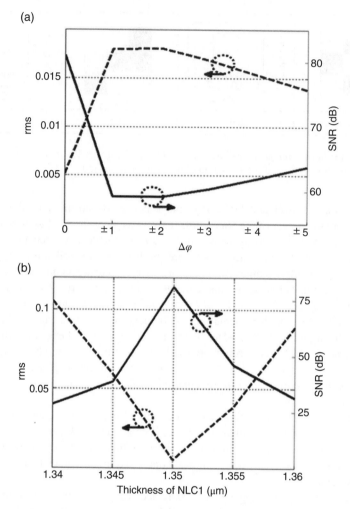

Figure 12.18 Tolerance in rms and $(SNR)_{rms}$ versus (a) the variations of φ and (b) the variations of the thickness using ψ (−90° and 90°)

12.4 Summary

In this chapter, an optical design for an image encryption scheme [20] which utilizes 3D PBG structures has been discussed. The optical encryption system has been modeled using the rigorous and accurate FITD method. In the optical encryption process, reflection with constant phase 180° over the frequency range 31.1–31.2 GHz is used to encrypt the pixels of the image. The use of the 180° phase has been demonstrated to enhance the security of the encoding system as the encryption and decryption keys are kept the same. Also, the demonstrated relatively wide bandwidth operation facilitates the correct sampling of the signal and signal recovery for intruders challenging. The reliability of the technique is evaluated by calculating the rms and the SNR_{rms} using the data of the source and retrieved images. The obtained test results rms=0.0759 and SNR_{rms}=6.426 reveal the good quality of the image retrieved by the suggested decryptor and prove the good overall performance of the design. Also, the noise added by encryption to the image has been quantified as the difference between the encrypted image power and the source image power. By normalizing the encryption noise power to the source image power, the SNR as good as $-3\,\text{dB}$ has been achieved. Therefore, the system is reasonably immune to external attack.

Further, another novel multiple symmetric image encryption system [21] has been introduced in detail. The optical design consists of a PBG block sandwiched between two layers of NLC of type E7. In the considered encryption system, two optical images have been encrypted simultaneously through modifying their phases by constant phase shifts $-90°$or $-180°$ over the wavelength range $1.377-1.3878\,\mu\text{m}$, respectively. The required phase shifts have been easily achieved by optimizing the thicknesses of the NLC layers. To retrieve the decrypted images, the same encryptor design can be used after modifying only the biasing state of the NLC_1 layer to obtain the complementary phase shift $90°$ and maintaining the unbiased NLC_2 layer as it is to obtain a phase shift of about $-180°$. The accuracy of the optical encryption technique has been quantified by calculating the values of the rms and the SNR_{rms} between the source and decrypted images. The estimated results rms$=0.0051$ and $\text{SNR}_{\text{rms}}=81\,\text{dB}$ for encryption using ψ ($-90°$ and $90°$) and rms$=0.0057$ and $\text{SNR}_{\text{rms}}=74\,\text{dB}$ for encryption using ψ ($-180°$) reflect the good overall performance of the optical encryption design. Moreover, the amounts of noise added by the optical encryption system were quantified to be as good as $-1.29\,\text{dB}$ for encryption using ($-90°$ and $90°$) phase shift and $6.658\,\text{dB}$ for encryption using ($-180°$) phase shift. Therefore, the optical encryption system [21] is reasonably insusceptible to outer attack. Further, it clearly has the advantage of offering the efficient use of the hardware to encrypt two images at the same time. Moreover, better security can also be achieved through a random jump from phase $-90°$ to $-180°$ over the same system. Extension of this idea is to encompass many phase shifts by using a bank of NLC layers at either side of the PBG block that can easily eliminate any external hacking.

References

[1] Simmon, G.J. (December 1979) Symmetric and asymmetric encryption. *J. ACM Comput. Surv.*, **11** (4), 305–330.

[2] Sarker, M.Z.H (2005) *A cost effective symmetric key cryptographic algorithm for small amount of data.* 9th International multitopic conference, IEEE INMIC, Karachi, December 24–25, 2005.

[3] Qin, W. and Peng, X. (2010) Asymmetric cryptosystem based on phase-truncated Fourier transforms. *Opt. Lett.*, **35**, 118–120.

[4] Rajput, S.K. and Nishchal, N.K. (2013) Image encryption using polarized light encoding and amplitude and phase truncation in the Fresnel domain. *Appl. Opt.*, **52** (18), 4343–4352.

[5] Karim, M.A., Islam, M.N., and Asari, V. (2013) Optics-based system for robust and reliable encryption. SPIE Newsroom, 31 May 2013.

[6] Li, Y., Kreske, K. and Rosen, J. (2000) Security and encryption optical systems based on a correlator with significant output images. *J. Appl. Opt.*, **39**, 5295–530.

[7] Tao, R., Xin, Y. and Wang, Y. (2007) Double image encryption based on random phase encoding in the fractional Fourier domain. *J. Opt. Express*, **15**, 16067–16077.

[8] Zhang, S. and Karim, M.A. (June 5, 1999) Color image encryption using double random phase encoding. *J. Microw. Opt. Technol. Lett.*, **21** (5), 318–322.

[9] Frauel, Y., Castro, A., Naughton, T.J. and Javidi, B. (August 2007) Resistance of the double random phase encryption against various attack. *Opt. Express*, **15** (16), 10253–10265.

[10] Carnicer, M., Usategui, S.A. and Juvells, I. (2005) Vulnerability to chosen-cyphertext attacks of optical encryption schemes based on double random phase keys. *Opt. Lett.*, **30**, 1644–1646.

[11] Naughton, T.J., Gopinathan, U., Monaghan, D.S. and Sheridan, J.T. (2006) A known-plaintext heuristic attack on the fourier plane encryption algorithm. *Opt. Express*, **14** (8), 3181–3186.

[12] Naughton, T.J., Hennelly, B.M. and Dowling, T. (2008) Introducing secure modes of operation for optical encryption. *J. Opt. Soc. Am. A*, **25**, 2608–2617.

[13] Clemente, P., Durán, V., Torres-Company, V., Tajahuerce, E. and Lancis, J. (2010) Optical encryption based on computational ghost imaging. *Opt. Lett.*, **35**, 2391–2393.

[14] Chen, W. and Chen, X. (2010) Space-based optical image encryption. *Opt. Express*, **18**, 27095–27104.

[15] Chen, W., Chen, X. and Sheppard, C.J.R. (2011) Optical double-image cryptography based on diffractive imaging with a laterally-translated phase grating. *Appl. Opt.*, **50**, 5750–5757.

[16] Pérez-Cabré, E., Cho, M. and Javidi, B. (2011) Information authentication using photon-counting double-random- phase encrypted images. *Opt. Lett.*, **36**, 22–24.

[17] Joannopoulos, J.D., Johnson, S.G., Winn, J.N. and Meade, R.D. (2007) *Photonic Crystals: Molding the Flow of Light*, 2nd edn, Princeton University Press, Princeton.

[18] Singh, M., Kumar, A., and Singh, K. (2009) *Encryption and decryption using a phase mask set consisting of a random phase mask and sinusoidal phase grating in the fourier plane*. International conference on optics and photonics, ICOP, 2009, Chandigarh, India, October 30, 2009.

[19] He, W., Peng, X. and Meng, X. (June 2012) Optical multiple-image hiding based on interference and grating modulation. *J. Opt.*, **14** (7), 75401–75410.

[20] Areed, N.F.F. and Obayya, S.S.A. (May 2012) Novel design of symmetric photonic bandgap based image encryption system. *J. Prog. Electromagn. Res.*, **30**, 225–239.

[21] Areed, N.F.F. and Obayya, S.S.A. (February 18, 2014) Multiple image encryption system based on nematic liquid photonic crystal layers. *J. Lightwave Technol.*, **32** (7), 1344–1350.

[22] Samra, A.S., Kishk, S.S. and Elnaggar, S.S. (June 2010) A compact lens-less optical image encoding system using diffraction grating. *Int. J. Comput. Sci. Netw. Secur.*, **10** (6), 141–145.

[23] Weiland, T., Ackermann, W., and Trommler, J. (2002) *Verfahren und anwendungen der feldsimulation*, Darmstadt.

[24] Krietenstein, B., Schuhmann, R., Thoma, P., and Weiland, T. (1998) *The perfect boundary approximation technique facing the challenge of high precision field computation*. Proceedings of the XIX International Linear Accelerator Conference (LINAC'98), pp. 860–862, Chicago, IL, 1998.

[25] Weiland, T. (1996) Time domain electromagnetic field computation with finite difference methods. *Int. J. Numer. Modell.*, **9**, 295–319.

[26] Canning, J. (2008) Fiber gratings and devices for sensors and lasers. *Laser Photon. Rev.*, **2** (4), 275–289.

[27] Prather, D.W., Sharkawy, A., Shi, S., Murakowski, J. and Schneider, G. (2009) *Photonic Crystals, Theory, Applications and Fabrication*, John Wiley & Sons, Inc., Hoboken.

[28] Gonzalez, R.C. and Wints, P. (1987) *Digital Image Processing*, 2nd edn, Addison-Wesley, Reading.

[29] Koshiba, M., Tsuji, Y. and Hikari, M. (January 2000) Time domain beam propagation method and its application to photonic crystal circuits. *IEEE J. Lightwave Technol.*, **18** (1), 102–110.

[30] Assefa, S., Rakich, P.T., Bienstman, P. *et al.* (December 20, 2004) Guiding 1.5 µm light in photonic crystals based on dielectric rods. *J. Appl. Phys. Lett.*, **85** (25), 6110–6112.

[31] Hameed, M.F.O. and Obayya, S.S. (2012) Modal analysis of a novel soft glass photonic crystal fiber with liquid crystal core. *J. Lightwave Technol.*, **30** (1), 96–102.

[32] Hameed, M.F.O. and Obayya, S.S. (2011) Polarization rotator based on soft glass photonic crystal fiber with liquid crystal core. *J. Lightwave Technol.*, **29** (18), 2725–2731.

[33] Cai, B. and Seeds, A.J. (1997) Optical frequency modulation links: theory and experiments. *IEEE Trans. Microw. Theory Tech.*, **45** (4), 505–511.

[34] Gan, X., Shiue, R., Gao, Y. *et al.* (2013) Chip-integrated ultrafast graphene photodetector with high responsivity. *Nat. Photon.*, **7**, 883–887.

[35] Patel, J.S. (1992) Electro-optic switch using a liquid crystal Fabry-Perot filter. *SPIE*, **1665**, 244–249.

[36] Lee, H. and Kim, J. (2007) Optical isolator using photonic crystal. US Patent 7,164,823 B216, January 2007.

[37] Taflove, A. and Hagness, S.C. (2005) *Computational Electrodynamics: The Finite-Difference Time-Domain Method*, 3rd edn, Artech House, Boston.

13

Optical Computing Devices Based on Photonic Liquid Crystal Layers

13.1 Introduction to Optical Computing

Recent studies have shown that the explosive growth in Internet traffic is demanding the next-generation Internet that can accommodate all the traffic in a cost-effective manner. This suggests the possibility of dynamically adapting the optical connections to carry these heavy flows. Currently, a packet has to traverse a certain number of routers before reaching its destination, and the network routers must analyze each packet and forward it in the direction of the destination node. Conventional electronic routers pose certain limitations on the full utilization of optical networks such as power handling issues, optoelectronic processes, and so on [1]. Therefore, the trend toward design of all-optical routers will help realize the future ultra-high-speed optical networks. Further, the use of photonic integrated circuits (PICs) in high-capacity optical networks has been researched, with experts claiming that all-optical packet switching and routing technologies hold promise to provide more efficient power and footprint scaling with increased router capacity. Nowadays, there is an emerging need for a photonic platform that will enable higher levels of flexible, functional, and cost-effective integration [2, 3]. The target is to integrate more and more optical processing elements into the same chip, and thus increase on-chip processing capability and system intelligence, while the merging of components and functionalities drives down packaging cost, bringing photonic devices one step (or more) closer to deployment in routing systems. There are many recent advances in the implementation of PIC for optical routing. These components include high-speed, high-performance integrated tunable wavelength converters and packet forwarding chips, integrated optical buffers, and integrated mode locked lasers [4–7]. To date, optical computers appear attractive for future broadband optical communication systems since they have some advantages, such as high speed, high throughput, and low power consumption. To build an optical computer, all-optical flexible signal processing devices are needed [8, 9]. Optical logic gates are considered as key elements in real-time optical processing and communication systems that perform the necessary functions at the nodes of the network,

Computational Liquid Crystal Photonics: Fundamentals, Modelling and Applications, First Edition.
Salah Obayya, Mohamed Farhat O. Hameed and Nihal F.F. Areed.
© 2016 John Wiley & Sons, Ltd. Published 2016 by John Wiley & Sons, Ltd.

such as data encoding and decoding, pattern matching, recognition, and various switching operations [10]. There are many different schemes that have been demonstrated to realize various all-optical logic gates, such as quantum dots [11], semiconductor optical amplifiers (SOAs) [12], multimode interference in SiGe/Si [13], and a nonlinear silicon-on-insulator (SOI) waveguide [14]. Despite the enormous technological progress, most suffer from certain limitations such as large size, difficulty in performing chip-scale integration, high power consumption, low speed, and spontaneous noise emission. Photonic crystals (PhCs), artificial periodic dielectric structures, have been designed to construct compact high-speed logic gates with dimensions of a few wavelengths of light being confined [15]. Due to their unique properties, they have many applications as couplers [16], drop filters [17], logic gates [18], optical routers [19], image encryption systems [20, 21], and dense chip-scale PICs [22].

In this chapter, five compact designs for an optical router, optical AND logic gate, optical OR logic gate, reconfigurable AND/OR optical logic gate, and optical storage have been presented and simulated using the finite-difference time domain (FDTD) technique.

First, a design of an easily and fully integrated Tbit/s optical router is presented in Section 13.2. The optical router consists of three photonic bandgap (PBG) waveguides with two nematic liquid crystal (NLC) layers [19]. The suggested device can be used to divert the light beam to one of the three PhC waveguides based on the biasing states of the two NLC layers. In this way, there are three different modes of operation used for routing data in the required direction. The optical router device offers crosstalk of 19 dB. In addition, it opens up the possibility of building multi-port optical routers through the use of a number of appropriately positioned NLC layers within the platform of the PBG structure.

Next, two novel designs for compact, linear, and all-optical OR and AND logic gates based on the PhC architecture have been introduced [23]. The optical gate devices are formed by the combination of the ring cavities and a Y-shaped line defect coupler placed between two waveguides. The performance of the presented logic gates has been analyzed and investigated using the 2D FDTD method. The considered design for the AND gate offers an ON to OFF logic-level contrast ratio of not less than 6 dB, and the design for the OR gate offers transmitted power of not less than 0.5. Additionally, the optical OR and AND logic gates can operate at bit rates (BRs) of around 0.5 and 0.208 Tbits/s, respectively. Further, the calculated fabrication tolerances of the optical logic gate devices show that the radii of the rods in the ring cavities need to be controlled with no more than ±10 and ±3% fabrication errors for optical OR and AND gates, respectively.

Then, we have introduced the fourth design for a compact, fully integrated, linear, flexible, and all-optical reconfigurable logic gate that has different modes of operation [24]. This tunable gate is built on a PhC platform and consists of three PhC waveguides with two silicon cylindrical cells filled with NLC of type E7. This design can operate as an AND gate and offers an ON to OFF logic-level contrast ratio of not less than 6 dB. Further, the same design can operate as an OR gate and offers transmitted power of not less than approximately 0.5. The optical logic gate can operate at a BR of around 100 Gbits/s.

Finally, a design of a compact, fully integrated, optical memory cell is presented. The optical memory cell is built on a PhC platform and consists of PhC waveguides with two silicon cylindrical cells filled with NLC of type E7. The designed optical storage can store and restore the optical signal based on the biasing states of the two NLC layers. It is expected that such designs will have the potential to be key components for future PICs as their simplicity and small size provides a step toward the next-generation fully integrated PhC circuit.

13.2 All-Optical Router Based on Photonic LC Layers

The scope of this section is to introduce an optical design for a photonic router with enhanced high levels of flexible integration and enhanced power processing [19]. In Ref. [19], a combination of PhC with a central cavity and NLC layers has been investigated, for the first time to the best of our knowledge, to realize an optical router. In the suggested design, the parameters of the PhC structure have been carefully selected in order to realize the PBG effect in the wavelength range of interest from 1.31 to 2 μm. This PBG effect allows isolation between the multiple PhC waveguides that are used to terminate different ports. Moreover, the central cavity within the PBG structure enhances the coupling between the three PhC waveguides. The routing to the required direction is mainly dependent upon the equivalent permittivity of this direction which can be easily obtained through the tuning of the NLC layers. The NLC have received much attention due to their external-field-dependent optical anisotropy and high birefringence [25]. This property of NLC allows a high degree of freedom in routing the light in the required direction. To demonstrate the excellent performance of the optical routing system, a crosstalk criterion has been adopted. The proposed router is a simple fabrication and has yet significantly efficient performance [19]. Therefore, the reported design provides a step toward the next generation of fully integrated photonic routers. In this study, a three-dimensional simulator based on the FDTD(3D-FDTD) algorithm was used [26] to simulate the optical router device.

This section is organized as follows. Following this introduction, a detailed description of the optical router will follow in Section 13.2.1. Section 13.2.2 presents the simulated design and the temporal response of the reported device. Section 13.2.3 discusses the fabrication tolerance for the optical router design.

13.2.1 Device Architecture

A schematic 3D and top view of the optical router device are shown in Figure 13.1a and b, respectively. The structure is made of a chip of PhC that has three PhC waveguides and two NLC layers. According to the electrical signals that excite the two NLC layers, different modes of operations for the NLCs can be obtained. These different modes of operations are used to focus the optical power coming from the input port (1) to one of the other two ports: (2) or (3).

13.2.1.1 PBG Structure

The PBG is composed of a 2D square lattice of dielectric silicon rods of refractive index 3.4 arranged in air [27], as shown in Figure 13.1. The diameter d of the silicon rod is fixed at $0.36a$, where a is the lattice constant which is taken as 0.58 μm. The parameters of the PhC structure shown in Figure 13.1 have been carefully selected in order to realize the PBG effect (zero transmission) in the wavelength range of interest (1.31–2 μm). This PBG effect allows isolation between the three PhC waveguides. By removing one central rod, a defect is formed in the considered PBG structure. The impeded cavity permits the flow of light with a wavelength that matches the wavelength of the defect mode through the PBG structure. Therefore, the central cavity enhances the coupling between the three PhC waveguides.

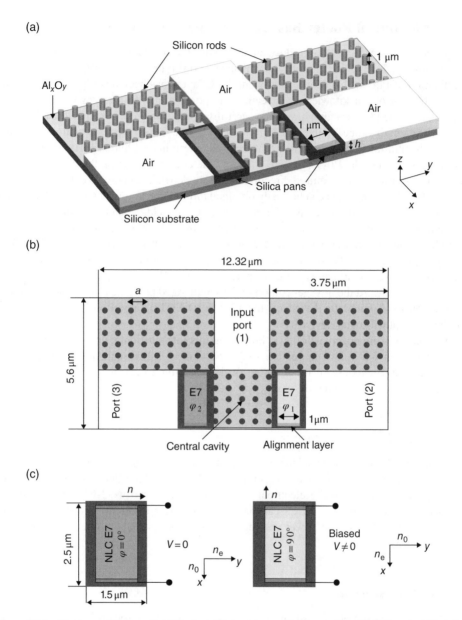

Figure 13.1 Router architecture: (a) 3D view, (b) top view, and (c) top views for LC molecular orientation for unbiased and biased states. *Source*: Ref. [19]

13.2.1.2 Liquid Crystals

LCs are fluids whose electrical permittivity is highly dependent on the orientation of their constituent molecules. A direct consequence of this ordering is the anisotropy of mechanical, electric, magnetic, and optical properties. LCs are very sensitive to an electric field, and this property allows their application to the design of various photonic devices. As shown in

Figure 13.1 and in order to build a practical device [28], two NLC layers of type E7 are placed in silica pans. The alignment layers of 100 nm polyimide are formed on the bottom surface of the silica walls along the x-axis. Two transparent electrodes are used to modify the biasing state of the NLC layer. The NLC of type E7 has been widely used in LC devices due to its large birefringence (~0.2) and wide nematic temperature range, from 10 to 59°C [29]. The relative permittivity tensor ε_r is previously defined in Eq. (12.3) [30].

Figure 13.1c shows top views of the LC molecular orientation for unbiased and biased states. In the unbiased state ($V_{LC} = 0$), the LC molecules progressively align along the y-axis with rotation angle $\varphi = 0°$, where φ represents the angle between the NLC director and y-axis. In this case, the light for both x- and z-polarizations is affected by the ordinary index n_o of the LC cell. However, in the biased state, the applied voltage across the biasing layers along the x-axis rotates the LC molecules. At saturation, the director is oriented perpendicular to the cell wall along the x-axis, with rotation angle $\varphi = 90°$. Therefore, the x-polarized light depends on the extraordinary index n_e of the NLC material.

13.2.1.3 System Operation

The system operation is based on the capability of routing an optical signal coming from port (1) depending on the biasing states of the two NLC layers at the left and right sides of the PBG layer. In this way, we will have three modes of operations. In the first mode of operation, the right NLC is biased, while the left NLC is not biased. In this case, the light will propagate through the PBG block from port (1) to port (2). In the second mode of operation, the right NLC is not biased, while the left NLC is biased. Therefore, the light will propagate through the PBG block from port (1) to port (3). Finally, in the third mode of operation, the left and right NLCs are not biased; then the light will be totally reflected back to the input port: port (1). This idea can be extended to encompass a router with many directions through the use of a bank of NLC layers at either side of the PBG block. This can be exploited to implement key functionalities required in the next-generation fully integrated photonic routers.

13.2.2 Simulation Results

In this study, the 3D-FDTD numerical method has been used to investigate the PBG of a 2D PBG structure shown in Figure 13.2a. Figure 13.2a shows a 2D array of high-index dielectric rods that reside on a silicon substrate isolated by a 200 nm layer of low index Al_xO_y, as reported in Ref. [31]. For the PhC, each rod has diameter 0.2088 μm and height 1 μm and is placed in a square lattice with a period of 0.58 μm. Figure 13.2b shows the input x-polarized mode which is located at the reference xy-plane [port (1)]. Because of the symmetrical nature of the device, only one-half of the computational domain ($x \times y \times z$) (0.5 μm × 3.86 μm× 10.528 μm) of the structure is analyzed. Figure 13.2c and d shows the transmission characteristics as a function of the wavelength across a 2D photonic crystal device with and without substrate, respectively. The creation of the PBG, from 1.3 to 2 μm, is observed in the case of no substrate while in the case of a device having isolated substrate, the PBG effect has been realized in the (1.48–1.53 μm) wavelength range of interest to the considered optical router design. For the rest of our simulations, the router built on a silicon substrate isolated by a layer of Al_xO_y has been adopted.

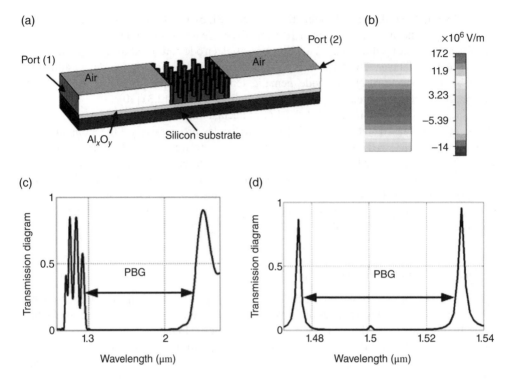

Figure 13.2 Photonic crystal platform: (a) 3D view, (b) input mode, (c) transmission diagram without substrate, and (d) transmission diagram with substrate. *Source*: Ref. [19]

The analysis of the optical router structure shown in Figure 13.1 for the first mode of operation with $\varphi_2 = 0°$ and $\varphi_1 = 90°$ is shown in Figure 13.3. In the simulation, each NLC layer is a cuboid structure with dimensions 2 μm × 1 μm × h. Both left and right silica pans should be of the same geometrical size, 2.5 μm × 1.5 μm × h with walls of thickness of 250 nm. Alignment layers of 100 nm polyimide are formed on the bottom surface of the silica walls along the x-axis. The wavelengths of the defect modes can be estimated by removing the NLC layers from the optical router and calculating the transmission characteristics S_{21} across the device. Figure 13.3a shows that the embedded cavity permits the flow of two optical waves with wide bandwidths centered at 1.495 and 1.512 μm, respectively. As observed from Figure 13.3b, the insertion of NLC layers with height $h = 0.3$ μm in the optical router shown in Figure 13.1 results in collision and scattering of the optical modes with the NLC layer, forming two groups of narrow scattered wavelengths (group 1 and group 2). Moreover, by increasing the thickness of the NLC layers to 0.7 μm, the two groups of the scattered notches move toward each other and tend to interfere, as shown in Figure 13.3c. By slightly increasing the NLC height to 0.8 μm, the interference of the scattered notches results in two strong sharp notches each of which will be routed only to port (2) which has the biased NLC layer, as shown in Figure 13.3d. It can be noted from Figure 13.3e that, by increasing the NLC height to 1 μm, the bandwidth of the light received by port (2) with NLC is much narrower than the bandwidth of the light received by port (2) without NLC, as shown in Figure 13.3a. Based on these results, the amount of light routed to either port (2) or port (3) is highly sensitive to the effective refractive

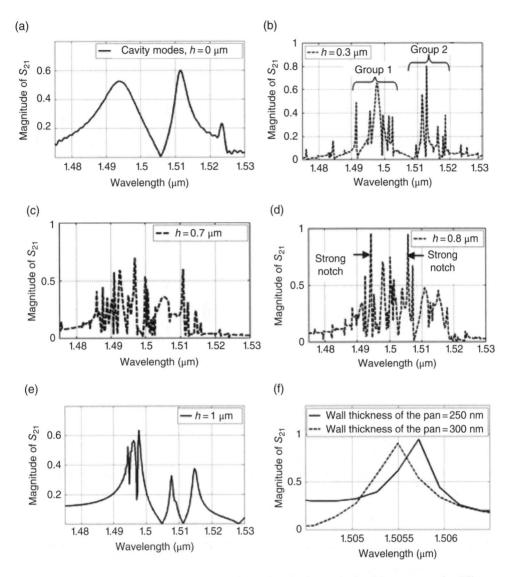

Figure 13.3 Analysis of optical router: (a) cavity modes, (b–e) magnitude of S_{21} parameter for different values of NLC height, and (f) magnitudes of S_{21} for $h = 0.8$ μm for different geometrical parameters of the silica pans

index of the corresponding waveguide of this port, which depends clearly on the height of the corresponding NLC layer and its biasing state. The transmission characteristics across the optical router using the optimum height $h = 0.8$ μm for different geometrical parameters of the silica pans are shown in Figure 13.3f. A 20% change in the pan thickness results in only a 0.013% change in the resonance wavelength. Nevertheless, our simulations have shown us that such a small shift will not affect the overall router performance. Figure 13.4a shows that by tuning to the first mode of operation, the majority of the incident wave will be totally routed

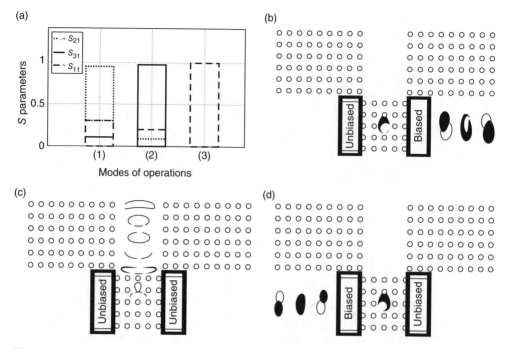

Figure 13.4 (a) *S*-parameters, (b–d) 2D steady-state field profile at $\lambda = 1.5057$ μm for the three different modes of operation for the optimum NLCs height of 0.8 μm. *Source*: Ref. [19]

to port (2) where the magnitudes of S_{21}, S_{31}, and S_{11} will be equal to 0.945, 0.111, and 0.305, respectively, at the resonance wavelength $\lambda = 1.5057$ μm. Alternatively, by tuning to the second mode of operation with $\varphi_2 = 90°$ and $\varphi_1 = 0°$ the magnitudes of S_{21}, S_{31}, and S_{11} will be equal to 0.0929, 0.97352, and 0.205, respectively. These values indicate that the majority of the optical wave with wavelength 1.5057 μm will be routed to port (3). Finally, by tuning to the third mode of operation with $\varphi_2 = \varphi_1 = 0°$ the magnitudes of S_{21}, S_{31}, and S_{11} will be 0.07, 0.08, and 0.9945, respectively. The numerical results show that the optical wave will be reflected back to port (1). It should be noted that the optical router device without a central defect at the optimum height 0.8 μm has difficulty in holding efficient coupling to the required direction, which demonstrates the importance of the central cavity in the suggested design. Figure 13.4b–d shows the steady state of the absolute electric fields along the *xy*-plane for the three different modes of operation at $\lambda = 1.5057$ μm for the optimum NLCs height (0.8 μm). The displayed field distributions for the three different modes of operations clearly agree with the behavior of the *S*-parameters shown in Figure 13.4a.

13.2.3 Fabrication Tolerance

The crosstalk performance of the optical router device is calculated as follows [32]:

$$\text{Cross talk} = 10 \log \left(\frac{P_{\text{on}}}{P_{\text{off}}} \right). \tag{13.1}$$

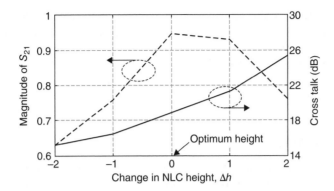

Figure 13.5 S_{21} magnitude and the crosstalk versus the change in NLC height Δh

Here, P_{on} is the power received by the active port in any of the three different modes of operation and P_{off} is the leaky power received by either of the two other ports. In this study, the device crosstalk reports a value around 19 dB in any case. The tolerance with respect to the error of fabrication has also been evaluated. During the fabrication process, errors can be made simultaneously on several parameters at a time, especially on the NLC height. Transmission parameter S_{21} and the crosstalk criterion versus the variation of the NLC height Δh have been evaluated and plotted, as shown in Figure 13.5. As may be observed from this figure, the overall acceptable tolerance will be within ±1 nm where there are acceptable crosstalk values within the range 16–22 dB with minimum percentage error for the transmission parameter S_{21} within the range 2–20%. The introduced LC-based optical router has potential applications for the design of tunable photonic devices, such as switches, filters, reconfigurable logic gates, optical memories, sensors, and so on.

13.3 Optical Logic Gates Based on Photonic LC Layers

The scope of this section is to introduce three novel designs of compact, linear, and all-optical OR, AND, and reconfigurable AND/OR logic gates based on a PhC platform [30, 31]. The salient features of the proposed device such as small device length, total avoidance of nonlinear optics, and straight PhC waveguide input arms enable easy optical integration. It is expected that such designs have the potential to be key components for future PICs due to their fast response, simplicity, small size, and flexible integration.

13.3.1 OR Logic Gate Based on PhC Platform

In this subsection, a design for an optical OR logic gate based on a 2D PhC platform is presented [30]. The structure is composed of two line defects, ring cavities and a Y-branch waveguide. In Ref. [30], the procedure for designing the complete structure is based on designing and optimizing half of it first. The physical dimensions of the half structure are appropriately optimized with the purpose of transmitting a signal with optimized peak from the input waveguide to the output waveguide. Next, two arms of the optimized half structure are merged in order to produce a complete logic gate. To evaluate the performance of the optical OR gate design, the transmission and the response period have been calculated. The optical OR gate design offers transmitted power of not less than 0.5, and it can operate at BRs of around 0.5 Tbits/s. Further, the calculated fabrication tolerance of the optical OR gate shows that the radii of the rods in the ring cavities need to be controlled with no more than ±10% fabrication error.

13.3.1.1 PhC Platform

The utilized 2D PhC platform in the implementation of the optical OR gate is shown in Figure 13.6 and consists of an array of size 49 × 40 of silicon rods arranged in a triangular lattice [23]. The refractive index and the diameter of the silicon rod and the lattice constant are chosen as 3.59, 0.2, and 0.54 μm, respectively. The PBG calculations depend on illuminating the 2D PhC structure with a Gaussian pulse from a homogeneous medium and investigating the propagation spectra around the 2D PhC structure [33]. Figure 13.6a shows the transmission spectra around the 2D PhC structure, for transverse magnetic (TM) and transverse electric (TE) polarization modes. This figure reveals that the considered PhC platform has incomplete PBG which only exists over the TM polarization mode and extend from 1.2165 to 1.9409 μm. Figure 13.6b shows the steady-state field profiles along the propagation xy-plane at $\lambda = 1.5$ μm for the TM and TE modes, respectively. The displayed field distributions clearly agree with the behavior of the transmission spectra where the light cannot pass for the TM mode but can pass through the PhC structure for the TE mode. To keep a PBG effect in the considered wavelength range of interest using a 3D PhC structure, the height of the PhC rods should be carefully selected. As indicated in Ref. [33], the optimum rod height should be around half a wavelength relative to an averaged index that depends on the polarization.

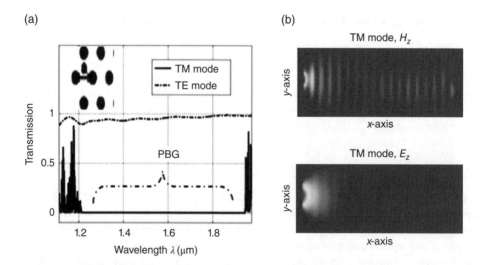

Figure 13.6 PhC structure: (a) transmission spectra and (b) field profiles

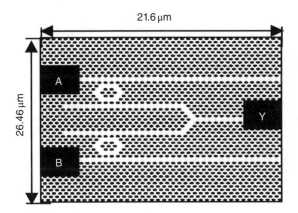

Figure 13.7   The optical OR gate structure. *Source*: Ref. [23]

Table 13.1   Truth table for OR logic gate

A	B	Logic output	Transmission $T = |E_o/E_{in}|^2$
0	0	0	0
0	1	1	≥ 0.5
1	0	1	≥ 0.5
1	1	1	≥ 0.5

13.3.1.2   Optical OR Gate Architecture

Figure 13.7 shows the schematic diagram of the optical OR logic gate using the 2D PhC platform shown in Figure 13.6a. The optical OR gate is formed by two line defects, two ring cavities and a Y-branch waveguide. The input signals come from the left ports of the upper and lower waveguides: A and B, respectively. The output signal is obtained from the right port of the middle waveguide, Y. It is expected that the use of a ring cavity permits a signal with a single peak to go from the input port to the output port. The design procedure is based on optimizing only the geometrical parameters of the upper half of the structure shown in Figure 13.7 in order to achieve maximum transmission [23].

Next is simulating the complete design shown in Figure 13.7 to mimic the truth table behavior of the OR logic gate shown in Table 13.1. It can be observed from the table that the output is logically "1" if any of the input values is 1. As indicated in Table 13.1, the transmission of the logical "1" is defined by 0.5 or more.

13.3.1.3   Results and Discussion for OR Gate

To verify the proposition in the previous section, a 2D-FDTD simulator [26] has been used to investigate and optimize the performance of the ring cavity and the Y-branch waveguide. First, the spectral response of the ring cavity is investigated as shown in Figure 13.8. In Figure 13.8a,

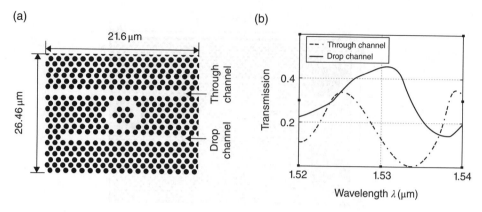

Figure 13.8   Simulation of ring cavity: (a) structure and (b) transmission

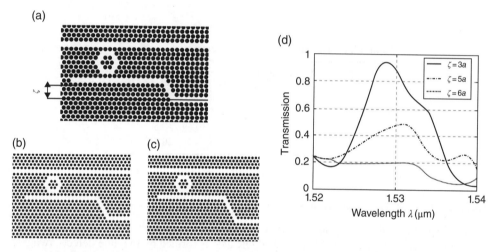

Figure 13.9   Bend with different ζ: (a) $\zeta = 3a$, (b) $\zeta = 5a$, (c) $\zeta = 6a$, and (d) transmission

the input signal comes from the left port of the upper waveguide, called the "through channel." The output signal is obtained from the right port of the lower waveguide called the "drop channel." Figure 13.8b shows the transmittance with respect to the wavelength for TM-like polarization of the incident light from the input port. It can be found from the figure that one peak with about 40% of the input power will be transmitted to the drop channel. The value of the dropped power can be controlled and optimized by bending the drop channel and taking the output depending on a different vertical length ζ as shown in Figure 13.9. Figure 13.9d shows the calculated dropped power of the structures shown in Figure 13.9a–c with ζ ranges from $3a$ to $6a$ for the wavelength range from 1.52 to 1.54 μm. It can be observed from this figure that, the dropped power increases with decreasing ζ and reaches its maximum value at $\zeta = 3a$.

Two arms of the optimized structure shown in Figure 13.9a are merged in order to produce a compact optical OR gate at the wavelength 1.529 μm, as shown in Figure 13.7. Figure 13.10a shows the steady-state field profiles at 1.529 μm for the optical OR gate for the last three cases shown in Table 13.1. Figure 13.10b shows the spectral transmission of the optical OR gate for

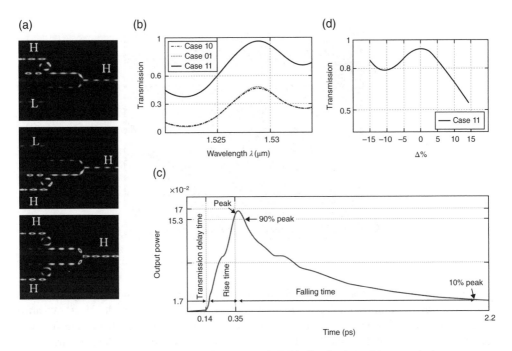

Figure 13.10 Results of the optical OR gate: (a) field distribution, (b) transmission spectrum, (c) time-evolving curve of the output power, and (d) fabrication tolerance

TM-like polarization of incident light from a single input port and both input ports. The figure shows that when the signal at $\lambda = 1.529$ μm is launched from port A, while the signal B is off it leads to forward-dropping (logic 1) at the output port Y where the transmission reads 0.5. The same procedure happens when signal B is on, while the signal A is off. Moreover, the excitation of the two input ports by two identical input signals results in forward-dropping (logic 1); the transmission reads 0.95. The BR of the optical OR gate has been determined using the time-evolving curve of the output power similar to Ref. [26]. Figure 13.10c shows that the time of the output power consists of three parts: one of which is due to transmission delay, that is, 0.14 ps, next is the time for the power to climb from 0.1% peak to 90% peak, 0.21 ps, and finally, the falling time from the 90% peak to the 10% peak, approximately 1.65 ps. Hence, the response period of the output power is 2 ps and the optical OR gate can operate at a BR of 0.5 Tbits/s. The tolerance with respect to the fabrication error has been evaluated. We have intentionally added $\pm\Delta\%$ tolerances in the radii of the central rods in the ring cavities. As indicated in Figure 13.10d, calculation of the fabrication tolerance of the suggested device found that the radii of the rods need to be controlled with no more than $\pm10\%$ fabrication error.

13.3.2 AND Logic Gate Based on a PhC Platform

The scope of this study is to introduce a compact design for an optical AND logic gate based on a 2D PhC platform [23]. The structure of the AND logic gate is composed of two line defects, ring cavities and a Y-branch waveguide. Here, the procedure for designing the complete structure is based on designing and optimizing one half first. The physical dimensions of

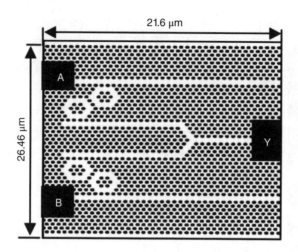

Figure 13.11 Optical AND gate structure. *Source*: Ref. [23]

the half structure are appropriately optimized with the purpose of transmitting a signal with an optimized peak or optimized null from input waveguide to output waveguide. Next, two arms of the optimized half structure are merged in order to produce a complete logic gate. To evaluate the performance of the optical AND gate design, the ON to OFF logic-level contrast ratio has been calculated. The AND gate design offers an ON to OFF logic-level contrast ratio of around 6 dB and it can operate at BRs of around 0.208 Tbits/s. The calculated fabrication tolerance of the AND gate design shows that the radii of the rods in the ring cavities need to be controlled with no more than ±3% fabrication error.

13.3.2.1 Optical AND Gate Architecture

Figure 13.11 shows the schematic diagram of the optical AND logic gate using the PhC platform shown in Figure 13.6. It can be seen that the AND gate design is formed by two line defects, four ring cavities and a Y-branch waveguide. The input signals come from the left ports of the upper and lower waveguides, A and B, respectively, while the output signal is obtained from the right port of the middle waveguide, Y. By the same criteria, the upper half of the AND structure will be optimized to control the magnitude of the transmittance. It is expected that using two ring cavities will result in dropping signal with multiple peaks and multiple nulls. Hence, the position and the magnitude of the null can be controlled through optimizing the geometrical parameters of the ring cavities and the Y-branch line defect to mimic the truth table behavior of the AND logic gate shown in Table 13.2. It can be observed from the table that the output is logically "1" if only both of the input values are 1.

13.3.2.2 Results and Discussion for AND Gate

The 2D FDTD simulator is also used to investigate and optimize the performance of the optical AND gate shown in Figure 13.11. First, the spectral response of the two coupled ring cavities is investigated as shown in Figure 13.12. As shown in Figure 13.12a, the input signal

Table 13.2   Truth table for AND logic gate

A	B	Logic output	$T = |E_o/E_{in}|^2$	Contrast ratio ≈ 6 dB
0	0	0	0	
0	1	0	≤ 0.4	
1	0	0	≤ 0.4	
1	1	1	≥ 0.8	

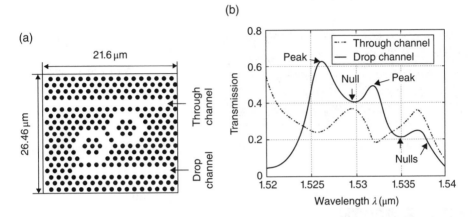

Figure 13.12   Simulation of PhC resonator: (a) structure and (b) transmission

comes from the left port of the upper waveguide called the "through channel." However, the output signal is obtained from the right port of the lower waveguide called the "drop channel." Figure 13.12b shows the transmittance with respect to the wavelength for TM-like polarization of the incident light from the input port. It can be found from the figure that the effect of two cavities results in transmission of multiple peaks and multiple nulls. The position of the null and its power magnitude can be controlled and optimized by bending the drop channel and taking the output depending on different vertical length ζ, as shown in Figure 13.13a–c. Figure 13.13d shows the calculated dropped power for the structure shown in Figure 13.13 with $\zeta = 3a$–$6a$. The positions of the nulls move to the left and the transmission value increases with increasing ζ.

A compact optical AND gate can be obtained at wavelength 1.538 μm by merging two copies of the optimized structure shown in Figure 13.13a as revealed from Figure 13.11. Figure 13.14a shows the steady-state field distributions at $\lambda = 1.538$ μm for the optical AND gate for the last three cases shown in Table 13.2. Figure 13.14b shows the contrast ratio of the optical AND gate for the incident light from a single input port and both input ports. When the signal at $\lambda = 1.538$ μm is launched at port A while signal B is off, it leads to reverse-dropping (logic 0) at the output port C where the transmittance reads 0.4. The same procedure happens when signal B is on while signal A is off. Moreover, the excitation of both ports A and B by two identical input signals results in forward-dropping (logic 1); the transmittance reads 0.8. From Figure 13.14c, it has been concluded that the response period of the output power is equal to 4.8 ps and the optical AND logic gate can operate at a BR of 0.208 Tbits/s.

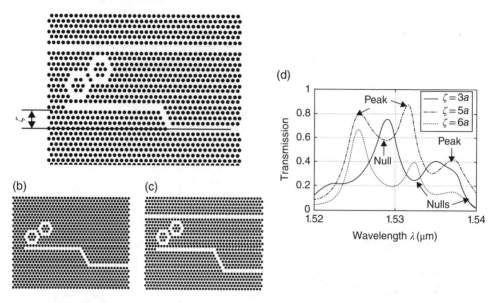

Figure 13.13 Bend with different ζ: (a) $\zeta = 3a$, (b) $\zeta = 5a$, (c) $\zeta = 6a$, and (d) transmission spectrum

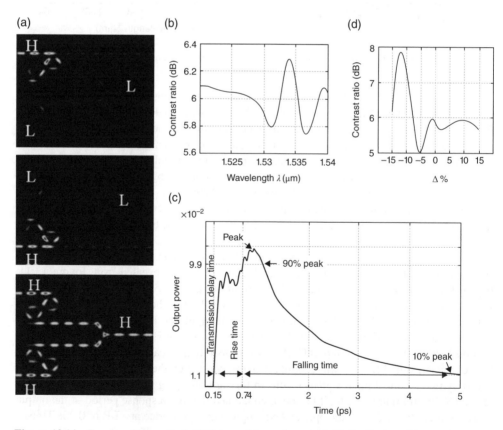

Figure 13.14 Results of the optical AND gate: (a) steady-state field distribution, (b) contrast ratios, (c) time-evolving curve of the output power, and (d) fabrication tolerance. *Source*: Ref. [23]

The tolerance with respect to the error of fabrication has been evaluated. We have intentionally added $\pm\Delta\%$ tolerance in the radii of the central rods in the ring cavities. As indicated in Figure 13.14d, the optimized central radii of the rod in the ring cavities need to be controlled with not more than $\pm 3\%$ fabrication error.

13.3.3 Reconfigurable Gate Based on Photonic NLC Layers

The scope of this section is to propose and simulate a compact reconfigurable optical logic gate that can be easily configured to operate as an AND or OR gate. The proposed optical gate is built around silicon on insulator photonic crystal platform and consists of three PhC waveguides with two silicon cylindrical cavities actuated by two NLC layers of type E7. A detailed description of the proposed gate will follow in Section 13.3.3.1. First, the procedure of the design is based on tuning the physical dimensions of some selected silicon rods in the proposed structure with the purpose of realization of OR operation. Next, the vertical displacement of the two NLC layers has been adjusted to allow reconfigurable gate (AND/OR) that can be easily configured based on biasing states of NLC layers. To evaluate the performance of the suggested design, steady-state field distributions, ON to OFF logic-level contrast ratio, transmission spectra, phase calculations, and the operation speeds have been calculated. The suggested reconfigurable gate offers an ON to OFF logic-level contrast ratio of around 6 dB and transmitted power of not less than 0.48 when configured to function as AND and OR gates, respectively. Further, the proposed reconfigurable gate handles high data rates of almost 100 Gbits/s. The use of the proposed reconfigurable optical logic gate instead of traditional PhC logic gate within the PhC platform can be applied for implementing the optical counterpart of field-programmable gate array (FPGA), and thereby obtaining programmable high-performance photonic circuit designs.

Following this introduction, the description of the suggested design and the simulation results have been demonstrated in the following subsections.

13.3.3.1 Device Architecture

Figure 13.15 shows the proposed reconfigurable gate that is built around the 2D PhC platform. It can be noted from the figure that the proposed two input reconfigurable gate is formed by three line defects and two silicon cavities infiltrated by NLC material of type E7. Further, the input beams come from the left ports of the upper and lower waveguides called "port (1)" and "port (2)," respectively. The output beam is obtained from the right port of the middle waveguide called "port (3)." It is expected that changing the biasing of the NLC layers from "no bias state" to "heavily biased state" enables the proposed structure to function as OR and AND gates, respectively. First, the design concepts of our proposal are based on tuning the geometrical parameters of some selected silicon rods shown in the squares denoted by S_2 and S_3 as illustrated in Figure 13.15. This is done to obtain maximum possible transmission, which is defined as $T = |E_o/E_{in}|^2$, where E_{in} is the input incident field either by port (1) or port (2) and E_o is the output received field by port (3). Next, we have adjusted the vertical displacement of the two NLC layers to obtain amenable transmission and realize reconfigurable AND/OR gate based on modifying the biasing states of the NLC layers. The two NLC layers are assumed initially at unbiased states ($\varphi = 0°$) where the direction of the molecules is aligned along the

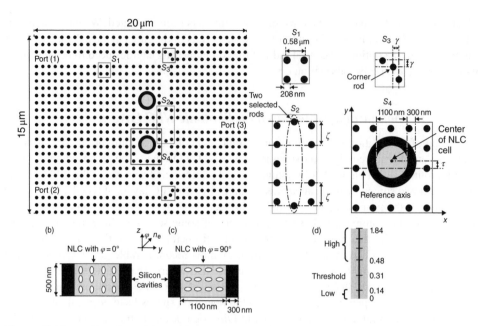

Figure 13.15 (a) Top view of the proposed optical reconfigurable gate, orientation of NLC molecules; (b) unbiased; (c) biased; and (d) transmission logic levels

Table 13.3 Truth table for the reconfigurable OR/AND logic gate

Input port (1)	Input port (2)	Logic output for OR gate Port (3)	Transmission $T = \lvert E_o/E_{in} \rvert^2$ $T = \lvert S_{31} \rvert^2$	Logic output for AND gate Port (3)	Transmission $T = \lvert E_o/E_{in} \rvert^2$ $T = \lvert S_{31} \rvert^2$
0	0	0	0	0	0
0	1	1	$1.84 \geq T \geq 0.48$	0	$T \leq 0.14$
1	0	1	$1.84 \geq T \geq 0.48$	0	$T \leq 0.14$
1	1	1	$1.84 \geq T \geq 0.48$	1	$1.84 \geq T \geq 0.48$

z-axis as shown in Figure 13.15b. The biasing states of the NLC layers can be modified from the unbiased states to heavily biased states ($\varphi = 90°$) by applying electric field along the y-axis. Figure 13.15c shows the NLC at heavily biased state (Figure 13.15) where the direction of the molecules is aligned along the y-axis. In the case of OR gate, it can be observed from Table 13.3 that the output is logically "1" if any of the input values is 1, while in the case of AND gate, the output is logically "1" if only both of the input values are 1. As indicated in Figure 13.15d and Table 13.3, the logic "1" is defined by transmission value that is between 0.48 and 1.84, while the logic "0" is defined by transmission value that is less than or equal to 0.14.

13.3.3.2 Bandgap Analysis of Photonic Crystal Platform

In this section, the 2D PhC of interest is designed using a silicon-on-insulator (SOI) platform, as illustrated in Figure 13.16a and b. It can be observed from the figure that the proposed PhC slab is designed in a commonly available SOI waver that is composed of 500 nm-thick silicon

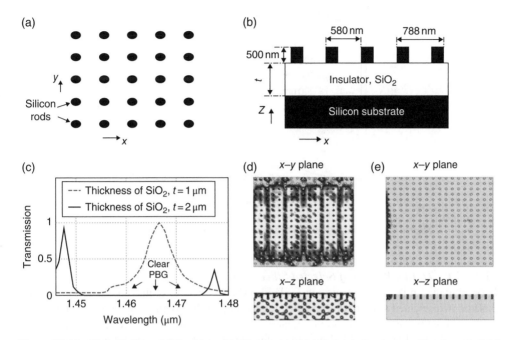

Figure 13.16 PhC platform: (a) Top view, (b) side view, and (c) transmission spectra. Steady-state field profiles for the PhC platform with 2 μm-thick SiO₂ at (d) $\lambda=1.447$ μm and (e) $\lambda=1.462$ μm.

layer ($n=3.4$) on top of silicon oxide layer ($n=1.45$) carried on a silicon substrate. In addition, the suggested PhC slab comprises cylindrical silicon rods that are arranged periodically in a square lattice with 500 nm rod height, 208 nm rod diameter, and 580 nm lattice spacing. 3D FDTD [34] method is used to calculate the PBG of the suggested PhC platform for TM polarization mode where the E-field is parallel to the dielectric rods. The suggested PhC platform has been numerically simulated to investigate the appropriate thickness, t of the buried oxide layer that decreases the leakage in the vertical direction and confines the mode within the silicon rods to achieve clear forbidden gap where no light propagation is allowed. Figure 13.16c shows the transmission characteristics as a function of the wavelength across the suggested 2D PhC platform at $t=1$ μm and $t=2$ μm. It can be indicated from the figure that PBG within the range 1.45–1.475 μm is realized at 2 μm-thick SiO₂. In addition to the transmission spectra, the steady-state field profiles along the xy and xz planes at $\lambda=1.447$ μm and $\lambda=1.462$ μm have been calculated for the PhC platform with 2 μm-thick SiO₂ as shown in Figure 13.16d and e, respectively. The field distributions clearly agree with the behavior of the transmission spectrum where the light cannot propagate along the structure at $\lambda=1.462$ μm while the light can pass at $\lambda=1.447$ μm through the structure.

13.3.3.3 Simulation Results of the Reconfigurable Gate

In this section, simulations based on 3D FDTD [34] have been performed with the purpose of tuning the dimensions of the proposed reconfigurable gate to function as AND and OR gates at two different NLC biasing states. The simulation strategy is based initially on showing the dimension tuning results of some selected silicon rods and the position tuning results of the NLC layers. Finally, the results of the proposed reconfigurable gate as AND gate and OR gate are presented.

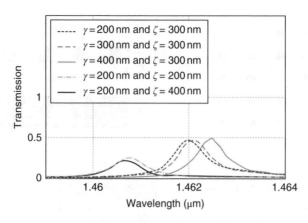

Figure 13.17 Tuning results of the proposed gate as a function of ζ and γ.

13.3.3.3.1   Tuning Results of Selected Silicon Rods

In order to maximize the transmission T, 3D FDTD is applied for tuning the vertical movement ζ of the two selected silicon rods and the movement γ of the up and down rods existing at the corners of the 90° bend shown in the squares denoted by S_2 and S_3, respectively, as illustrated in Figure 13.15. Figure 13.17 shows the calculated transmission T as a function of γ and ζ when the feeding comes only from port (1) and the NLCs are unbiased. The figure shows that the features of the resonant wavelength received by port (3) can be controlled by varying the values of γ and ζ. Further, at $\gamma=400\,\mathrm{nm}$ and $\zeta=300\,\mathrm{nm}$, T reaches its maximum value of about 0.5 at $\lambda=1.4624\,\mu\mathrm{m}$.

13.3.3.3.2   Tuning Results of NLCs

In this subsection, FDTD simulator [34] is applied for tuning the vertical displacement, τ of the two NLC cells as shown by the square labeled S_4 in Figure 13.15. Figure 13.18a shows the calculated transmission T at $\gamma=400\,\mathrm{nm}$, $\zeta=300\,\mathrm{nm}$, and $\tau=0\,\mathrm{nm}$ with unbiased and heavily biased NLC cells. It can be noted from the figure that varying the biasing state of the NLC cells shifts the transmission peak from $\lambda=1.4624\,\mu\mathrm{m}$ to a higher wavelength, $\lambda=1.46292\,\mu\mathrm{m}$, thereby generating two different transmission levels, $T=0.5$ and $T=0.27$, at $\lambda=1.4624\,\mu\mathrm{m}$. It is expected that changing the vertical displacement can be helpful for maximizing the difference between the two generated transmission levels and obtaining the high and low transmission levels previously defined in Table 13.3 at the unbiased and heavily biased NLC cells. Figure 13.18b and c shows the obtained results at the displacements $\tau=100\,\mathrm{nm}$ and $\tau=-100\,\mathrm{nm}$, respectively. It can be observed from Figure 13.18c that by moving the NLC cells downward, $\tau=-100\,\mathrm{nm}$, the transmission levels at $\lambda=1.462\,\mu\mathrm{m}$ reach a maximum value (high ≥ 0.48) and a minimum value (low ≤ 0.14) at unbiased and biased NLC cells, respectively.

13.3.3.3.3   Reconfigurable Gate as OR Gate

According to the obtained tuning results shown in Figure 13.18c, the proposed reconfigurable gate with tuned dimensions, $\gamma=400\,\mathrm{nm}$, $\zeta=300\,\mathrm{nm}$, and $\tau=-100\,\mathrm{nm}$, and with unbiased NLC cells can be used to mimic the operation of optical OR gate defined in Table 13.3. Figure 13.19a shows the calculated transmission spectra for the reconfigurable gate as OR gate. In addition to the transmission calculations, the switching capabilities of the proposed gate can be more

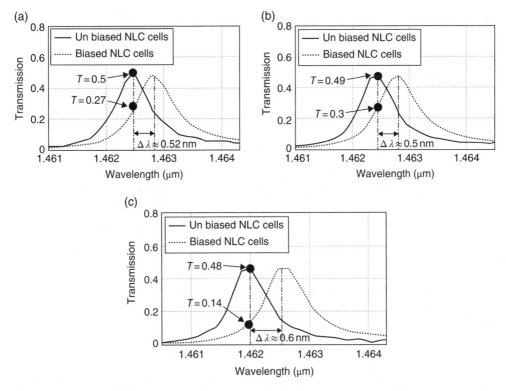

Figure 13.18 Tuning results of the proposed gate as a function of the vertical displacement, *T*.

evident by the calculated steady-state field profiles at $\lambda = 1.462\,\mu m$ for the last three input cases shown in Table 13.3, as illustrated in Figure 13.19b. The figure shows that when the beam at $1.462\,\mu m$ is launched at port (1), while the beam at port (2) is off, it leads to the forward-dropping (logic 1) at the output port (3) where the transmission reads 0.48. The same procedure happens when beam at port (2) is on, while the beam at port (1) is off. Moreover, the excitation of the two input ports by two identical input beams results in to the forward-dropping (logic 1); the transmission reads 1.84. Further, the Figure 13.19a shows that taking 0.31 level as a threshold between the high and low levels (defined previously in Table 13.3), the OR gate performs probably at $\lambda = 1.462\,\mu m$ with about 0.8 nm bandwidth. The high transmission value that is obtained in the last case 11 shown in Figure 13.19b can be explained by calculating the phase difference between the pair of beams received by port (3). The phase difference can be easily investigated by calculating the phases of the scattering parameters S_{31} and S_{32}. It can be noted from Figure 13.19c that the phases of the S_{31} and S_{32} are completely identical over the wavelength range of interest, and this can be explained as the traveling paths inside the proposed structure for both beams are identical. This implies that the phase difference between the two beams is equal to zero; thereby, the two beams interfere constructively on the output side, and as a result maximum transmission has been obtained. The bit rate (BR) of the optical gate can be calculated as the reciprocal of the response period that represents the duration starting from zero up to time at which the normalized power reaches 0.1. Figure 13.19d shows that the response period of the output power is equal to about 9.9 ps and the proposed reconfigurable gate as OR gate handles BR of 101 Gbits/s.

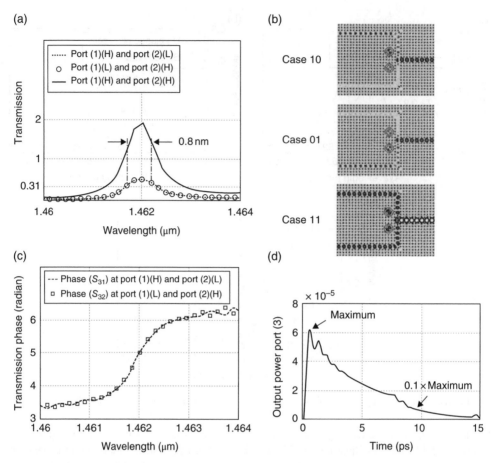

Figure 13.19 Proposed reconfigurable optical logic gate as OR gate: (a) Transmission spectra, (b) steady-state field distribution at $\lambda = 1.462\,\mu m$, (c) phases of S_{31} and S_{32}, and (d) time-evolving curve of the output power.

13.3.3.3.4   Reconfigurable Gate as AND Gate

To mimic the operation of optical AND gate defined in Table 13.3, the proposed reconfigurable gate with the same tuned dimensions, $\gamma = 400\,nm$, $\zeta = 300\,nm$, and $\tau = -100\,nm$ and heavily biased NLC cells can be used. Figure 13.20a shows the calculated contrast ratio in decibels for the reconfigurable gate as AND gate. The contrast ratio is defined as the ratio between the calculated transmission when the feeding comes from both input ports and the calculated transmission when the feeding comes from single input port. The figure shows that the contrast ratio is about 6 dB at $\lambda = 1.462\,\mu m$ with a wide bandwidth. This can be explained as, when the beam at $\lambda = 1.462\,\mu m$ is launched at port (1) while the beam at port (2) is off, it leads to the reverse-dropping (logic 0) at the output port (3) where the transmittance reads 0.14. The same procedure happens when beam at port (2) is on, while the beam at port (1) is off. Moreover, the excitation of both ports (1) and (2) by two identical input

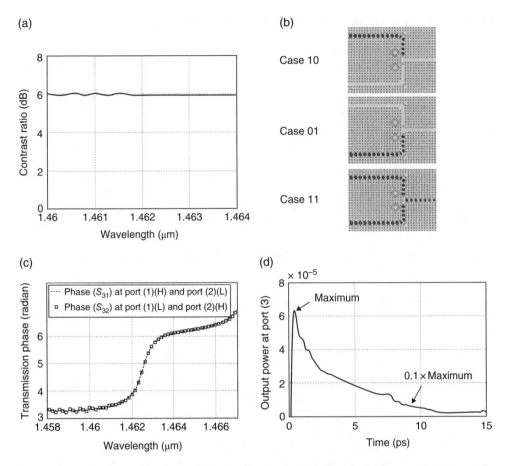

Figure 13.20 Results of the proposed reconfigurable optical logic gate as AND gate: (a) Contrast ratio, (b) steady-state field profiles at $\lambda = 1.462\,\mu m$, (c) phases of scattering parameters S_{31} and S_{32}, and (d) time-evolving curve of the output power.

beams leads to the forward-dropping (logic 1); the transmission reads 0.56. In addition, the switching capabilities of the proposed gate can be more evident by the calculated steady-state field profiles at $\lambda = 1.462\,\mu m$ for the last three input cases shown in Table 13.3, as illustrated in Figure 13.20b. The phase difference between the pair of beams received by port (3) in the last case 11 shown in Figure 13.20b can be investigated from the calculated phases of the scattering parameters S_{31} and S_{32} shown in Figure 13.20c. The figure shows that the phases of S_{31} and S_{32} are completely identical over the wavelength range of interest, and this implies that the phase difference between the two beams is equal to zero; thereby, the two beams interfere constructively on the output side, and as a result maximum transmission has been obtained. Figure 13.20d shows that the response period of the output power is equal to about 8.9 ps and the proposed reconfigurable gate as AND gate can operate at a BR of 112 Gbits/s.

13.4   Optical Memory Based on Photonic LC Layers

Information storage technology has always been under development, starting from the hard disk drive to Blu-ray, nanophotonic optical storage [35] and recently the pasmonic optical storage [36]. Recently, optical memory represents an ideal solution for storing large quantities of data very inexpensively. While it is not commonly practical for use in computer processing, there are successful third generations of compact discs (CDs), DVDs, and Blue-ray [37]. Around the 2000s, a search started for searching for a new generation for optical storage which can be introduced as a fourth generation. The fourth generation comprises different technologies, such as multilayer recording technology, holographic data storage technology (HODS), and near-field plasmonics for optical storage (NFPOS). In the multilayer recording technology, multiple semitransparent recording layers are used to increase the storage capacity. There are some commercial products that use this technology, such as Blu-ray disc that uses dual layers and can store up to 50 GB and triple layers that can store up to 128 GB. Nowadays, the developing target is to add multilayers up to 16 recording semitransparent layers to reach 512 GB [37]. Further, HODS technology is a promising technology and is based on utilizing two beams of light—a signal beam and a reference beam—to record data. The data to be recorded are modulated by a 3D code and then spotted onto the recordable medium. The recordable medium is irradiated by the reference beam, and the resultant 3D interference fringes from the two overlaid beams are recorded as data. This technology has a difficulty in controlling the relative distance between the objective lens [38]. Nowadays, NFPOS technology that utilizes metallic nanostructures gains much interest among researchers. The storage mechanism of this technology is based on subjecting several nano-holes etched in a metal sheet to light pulses. The characteristics of the stored beam vary from one design to another depending on its geometry, materials, and dimensions. The main drawbacks of NFPOS technology are the possibility of mechanical failure of the rotating disc and high crosstalk between two adjacent storage cells [39].

The scope of this section is to introduce a design for optical storage with a high level of flexible integration and enhanced power processing. Figure 13.21a shows the basic idea of the presented storage in storing and restoring the optical signal. As indicated from the figure, the steps for storing and restoring the optical signal can be summarized as follows:

1. Set the right switch opened and close the left switch, this permits the light beam to flow toward the PhC waveguide.
2. Instantaneously open only the left switch, this action should be fast enough to prevent the light beam reflecting and escaping from the cavity generated between the two opened switches.
3. The stored optical signal can be restored from the right port only by closing the right switch.

Figure 13.21b shows a schematic diagram for the implemented optical memory using a PhC platform and NLC cells. It can be observed from the figure that the device is composed of a PhC waveguide (cavity) surrounded by two NLC cells. The two NLC cells are used to perform the operations of the switches suggested by the basic idea shown in Figure 13.21a.

This section is organized as follows. Following this introduction, a detailed description of the PhC platform will follow in Section 13.4.1. Section 13.4.2 presents the simulated design and the temporal response of a tunable switch. Section 13.4.3 introduces the simulation results of the optical memory. Finally, the fabrication challenges will be discussed in Section 13.4.4.

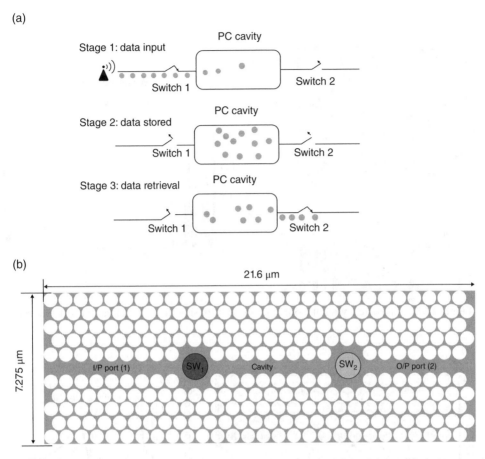

Figure 13.21 Optical memory: (a) suggested steps for storing and restoring optical signal and (b) implemented optical memory using photonic crystal platform and NLC cells

13.4.1 PhC Platform

The 2D PhC platform used to implement the optical storage is shown in Figure 13.22 and consists of an array of 49 × 40 air holes arranged in a triangular lattice and etched in a silicon substrate. The refractive index of the substrate, the diameter of the air hole, and the lattice constant are chosen as 3.59, 0.705, and 0.75 µm, respectively. The PBG calculations depend on illuminating the 2D PhC structure with a Gaussian pulse from a homogeneous medium and investigating the propagation spectra around the 3D structure. Figure 13.22c shows the calculated transmission spectrum around the 3D structure. This reveals that the considered PhC platform has a complete PBG from $\lambda = 1.4$ µm to $\lambda = 1.875$ µm.

13.4.2 Tunable Switch

In this subsection, the design of the optical switch using a combination of a PhC and LC will be discussed. As shown in Figure 13.23a, only one hole of diameter 1000 nm is infiltrated with NLC of type E7 and is inserted at the middle of the PhC waveguide. According to the biasing

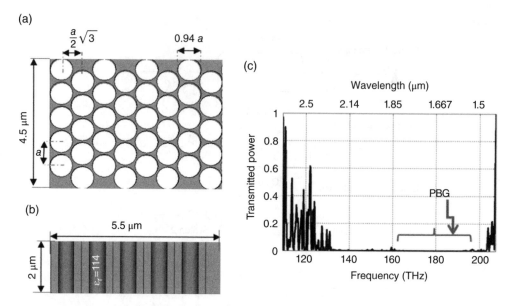

Figure 13.22 Photonic crystal platform: (a) top view, (b) side view, and (c) transmission diagram

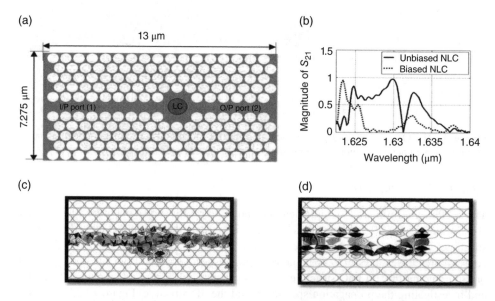

Figure 13.23 Tunable switch: (a) structure top view, (b) scattering parameter S_{21}. Steady-state field distribution at $\lambda = 1.63$ for (c) unbiased NLC and (d) Biased NLC

state of the NLC, the light is allowed to or prohibited from propagating through the waveguide from the input port to the output port. For example, in the biased state no light propagates (OFF state). However, in the unbiased state the light can propagate through the switch (ON state). Figure 13.23b shows the simulated transmission characteristics for the suggested

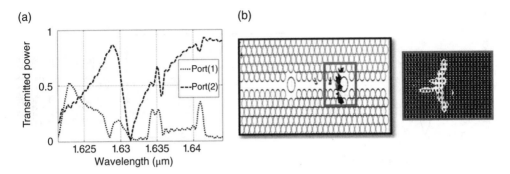

Figure 13.24 Storing optical signal in optical memory: (a) transmitted power and (b) steady-state field distribution at $\lambda = 1.6313$ μm

switch. It can seen that at the operating wavelength $\lambda = 1.63$ μm, for the biased state ($\varphi = 90°$, OFF state) $S_{21} = 0.1$ which means that no optical wave is transmitted to the output port from the input port. However, for the unbiased state ($\varphi = 0°$, ON state) $S_{21} = 0.95$, which means that the optical wave can be transmitted to the output port from the input port. In addition, Figure 13.23c and d shows the steady-state field distribution for ON and OFF states, respectively. It is clear from these figures that approximately the power will propagate for the ON state (Figure 13.23c) and vice versa for the OFF state (Figure 13.23d).

13.4.3 Simulation Results

In this section, the 3D-FDTD simulator [34] is applied to numerically simulate the optical memory shown in Figure 13.22b. Figure 13.24a shows the calculated transmitted power with two biased NLC cells. It can seen that the value of the transmitted power to either port (1) or port (2) is nearly zero at $\lambda = 1.6313$ μm which means that the signal is confined between the two NLC cells and stored in the middle PhC waveguide. This behavior is validated by calculating the steady-state field distribution at the wavelength 1.6313 μm, as shown in Figure 13.24b. It can seen from the figure that there is a good agreement with the transmission characteristics shown in Figure 13.24a where the fields are totally confined within the middle PhC waveguide. Figure 13.25a shows the calculated transmitted power with biased left NLC cell and unbiased right NLC cell. It can seen that the value of the transmitted power to port (2) is nearly 0.8 at $\lambda = 1.6313$ μm, which means that the signal is restored from the middle PhC waveguide. This behavior is validated by calculating the steady-state field distribution at the wavelength 1.6313 μm, as shown in Figure 13.25b.

13.4.4 Fabrication Challenges

As reported in Ref. [40], the dynamical optical response of the NLC after applying a voltage across the cell is approaching 10 ms for transmissive liquid crystal displays (LCDs) and less than 1 ms for color sequential micro-displays. However, for the optical memory shown in Figure 13.21, the required response time should be around orders of femto seconds. This presents a very technical complex challenge to nematic optical memories. One solution for

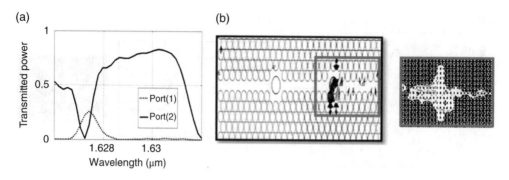

Figure 13.25 Restoring optical signal from optical memory: (a) transmitted power and (b) steady-state field distribution at $\lambda = 1.6313$ μm

meeting the challenge is to find appropriate fast response LC materials. As indicated by Wang [40], the response time of nematic LC materials can be improved either by constructing new molecular structures with low rotational viscosity or developing high birefringence. Another solution for meeting this challenge is to design an appropriate PhC isolator that permits propagation in only one direction. By following the left switch used in the optical memory shown in Figure 13.21 by a PhC isolator, backward propagation will be prohibited and hence the problem of low time response of the LC cell will vanish.

13.5 Summary

In this chapter, five compact designs for an optical router, optical AND gate, optical OR gate, reconfigurable optical AND/OR logic gate, and optical storage have been presented and numerically simulated using FDTD. Most of the considered designs are based on the combination of the PhC and LC layers.

First, a novel fully integrated optical router is presented and analyzed using an accurate 3D-FDTD method. The optical router design consists of a PBG platform with three PhC waveguides and two NLC layers. The operation of the router is based on the capability of routing the optical signal to any of the three PhC waveguides, depending on three different modes of operation. Each mode of operation represents a couple of biasing states at either $\varphi = 0°$ or $90°$ for the two NLC layers. The simulation results show a crosstalk of about 19 dB with a compact and simple structure, designed to operate with an optical range and compatible with PICs. The optical router can be designed with many directions by using a bank of NLC layers at either side of the PBG block.

After that, two optical logic gates have been analyzed and numerically simulated using the 2D-FDTD method. The optical design for the AND gate offers an ON to OFF logic-level contrast ratio of not less than 6 dB and the optical design for the OR gate offers transmitted power of not less than 0.5. Moreover, the optical OR and AND logic gates can operate at BRs of around 0.5 and 0.208 Tbits/s, respectively. We have also shown the fabrication tolerances of the optical gates and found that the radii of the central rod in ring cavities need to be controlled with no more than ±10 and ±3% fabrication errors for optical OR and AND gates, respectively.

Next, we have introduced a fourth design for a compact, linear, flexible, and tunable optical logic gate with different modes of operation. The 2D planar photonic crystal platform is composed of square array of silicon rods on top of 2 μm-thick silicon oxide layer carried on a silicon substrate. The proposed reconfigurable gate performs the operation of OR gate with transmission of not less than 0.48, operation speed of 101 Gbits/s, and bandwidth of about 0.8 nm. In addition, the same gate with modified NLC biasing performs the operation of AND gate with ON to OFF logic-level contrast ratio and operation speed of 6 dB and 112 Gbits/s, respectively. The proposed reconfigurable optical gate can be used as a key component for future programmable PICs.

Finally, a design of a compact, fully integrated, optical memory storage is presented. The optical memory cell is built on a PhC platform and consists of PhC waveguides with two silicon cylindrical cells filled with two NLC of type E7. The design can store and restore the optical signal based on the biasing states of the two NLC layers.

It is expected that the designs presented in this chapter have the potential to be key components for future PICs due to their simplicity and small size. This will provide a step toward the next-generation fully integrated PhC circuit.

References

[1] Yang, X. and Ramamurthy, B. (March 2005) Dynamic routing in translucent WDM optical networks: the intra-domain case. *J. Lightwave Technol.*, **23** (3), 955–971.

[2] Nagarajan, R., Kato, M., Lambert, D. *et al.* (2012) Coherent, superchannel, terabit/s InP photonic integrated circuits. *J. Semicond. Sci. Technol.*, **27**, 094003.

[3] Lipson, M. (December 2005) Guiding, modulating, and emitting light on silicon-challenges and opportunities. *J. Lightwave Technol.*, **23** (12), 4222–4238.

[4] Lipson, M. (November 2006) Compact electro-optic modulators on a silicon chip. *IEEE J. Sel. Top. Quantum Electron.*, **12** (6), 1520–1526.

[5] Liu, A. (2007) Announcing the world's first 40G silicon laser modulator. Intel.

[6] Xia, F., Rooks, M., Sekaric, L. and Lasov, Y. (September 2007) Ultra-compact high order ring resonator filters using submicron silicon photonic wires for on-chip optical interconnects. *J. Opt. Express*, **15** (19), 11934–11941.

[7] Xu, Q., Schmidt, B., Pradhan, S. and Lipson, M. (May 2005) Micrometre-scale silicon electro-opticmodulator. *Nature*, **435** (7040), 325–327.

[8] Mukherjee, B. (October 2000) WDM optical communication networks: progress and challenges. *IEEE J. Sel. Areas Commun.*, **18** (10), 1810–1824.

[9] Abdeldayem, H. and Frazier, D.A. (September 2007) Optical computing: need and challenge. *Commun. ACM*, **50** (9), 60–62.

[10] Baoa, J., Xiaoa, J., Fana, L. *et al.* (October 2014) All-optical NOR and NAND gates based on photonic crystal ring resonator. *J. Opt. Commun.*, **329** (15), 109–112.

[11] Pooley, M. (2012) Controlled-NOT gate operating with single photons. *Appl. Phys. Lett.*, **100** (21), 211103.

[12] Stubkjaer, K.E. (2000) Semiconductor optical amplifier-based all optical gates for high-speed optical processing. *IEEE J. Sel. Top. Quantum Electron.*, **6** (6), 1428–1435.

[13] Li, Z., Chen, Z. and Li, B. (February 2005) Optical pulse controlled all-optical logic gates in SiGe/Si multimode interference. *J. Opt. Express*, **13** (3), 1033–1038.

[14] Fujisawa, T. and Koshiba, M. (2006) All-optical logic gates based on nonlinear slot-waveguide couplers. *J. Opt. Soc. Am. B*, **23** (4), 684–691.

[15] Johnson, S. G. and Joannopoulos, J. D. (February 3, 2003), *Introduction to Photonic Crystals: Bloch's Theorem, Band Diagrams, and Gaps (But No Defects)*.

[16] Liu, C.-Y. and Chen, L.-W. (July 2005) The analysis of interaction region of elliptical pillars of a directional photonic crystal waveguide coupler. *Phys. E*, **28** (2), 185–190.

[17] Rashki, Z., Mansouri, M.A. and Rakhshani, M.R. (2013) New design of optical add-drop filter based on triangular lattice photonic crystal ring resonator. *J. Appl. Basic Sci.*, **4** (4), 985–989.

[18] Rani, P., Kalra, Y. and Sinha, R.K. (February 2013) Realization of AND gate in Y shaped photonic crystal wave-guide. *J. Opt. Commun.*, **298–299**, 227–231.

[19] Areed, N.F.F. and Obayya, S.S.A. (June 2013) Novel all-optical liquid photonic crystal router. *IEEE Photon. Technol. Lett.*, **25** (13), 1254–1257.

[20] Areed, N.F.F. and Obayya, S.S.A. (May 2012) Novel design of symmetric photonic bandgap based image encryption system. *Prog. Electromagn. Res.*, **30**, 225–239.

[21] Areed, N.F.F. and Obayya, S.S.A. (January 2014) Multiple image encryption system based on nematic liquid photonic crystal layers. *J. Lightwave Technol.*, **32** (7), 1344–1350.

[22] Dash, S. and Tripathy, S.K. (November 2012) Y-shaped design in two dimensional photonic crystal structure for applications in integrated photonic circuits. *J. Light Electron Opt.*, **124** (18), 3649–3650.

[23] Younis, R.M., Areed, N.F.F. and Obayya, S.S.A. (July 2014) Fully integrated AND and OR optical logic gates. *IEEE Photon. Technol. Lett.*, **26** (19), 1900–1903.

[24] Areed, N. F. F. and Obayya, S. S. A. (2016) Reconfigurable photonic crystal logic gates, to be submitted to *IEEE Photon. Technol. Lett.*

[25] Hameed, M.F.O. and Obayya, S.S. (2012) Modal analysis of a novel soft glass photonic crystal fiber with liquid crystal core. *J. Lightwave Technol.*, **30** (1), 96–102.

[26] Obayya, S. (January 2007) Improved complex envelope alternative direction implicit finite difference time domain method for photonic bandgap cavities. *J. Lightwave Technol.*, **25** (1), 440–447.

[27] Koshiba, M., Tsuji, Y. and Hikari, M. (January 2000) Time domain beam propagation method and its application to photonic crystal circuits. *IEEE J. Lightwave Technol.*, **18** (1), 102–110.

[28] Dickie, R., Baine, P., Cahill, R. *et al.* (May 24, 2011) Electrical characterisation of liquid crystals at millimetre wavelengths using frequency selective surfaces. *Electron. Lett.*, **48** (11), 611–612.

[29] Park, H., Parrott, E.P.J., Fan, F. *et al.* (2012) Evaluating liquid crystal properties for use in terahertz devices. *J. Opt. Express*, **20** (11), 11899.

[30] Ren, G., Shum, P., Yu, X., Hu, J., Wang, G. and Gong, Y. (2008) Polarization dependent guiding in liquid crystal filled photonic crystal fibers. *Opt. Commun.*, **281** (15), 1598–1606.

[31] Assefa, S., Rakich, P.T., Bienstman, P. *et al.* (December 20, 2004) Guiding 1.5 μm light in photonic crystals based on dielectric rods. *J. Appl. Phys. Lett.*, **85** (25), 6110–6112.

[32] Vázquez, C., Pena, J.M.S., Vargas, S.E., Aranda, A.L. and Perez, I. (August 2003) Router optical fiber sensor networks based on a liquid crystal cell. *IEEE Sens. J.*, **3** (4), 513–518.

[33] Jing, C., Jiyu, T., Peide, H. and Junfang, C. (April 2009) Optical properties in 1D photonic crystal structure using Si/C60 multilayers. *J. Semicond.*, **30** (4), 043001.

[34] Lumerical, www.lumerical.com. (accessed January 5, 2016).

[35] Gu, M., Li, X. and Cao, Y. (2014) Optical storage arrays: a perspective for future big data storage. *Light Sci. Appl.*, **3**, e177.

[36] Elrabiaey, M.A., Areed, N.F.F., and Obayya, S.S.A. (2015) Plasmonic Optical Binary Storage Based on Nematic Liquid Crystal Layers. The 36th PIERS. Prague, Czech Republic, 6–9 July, p. 336.

[37] Inoue, M., Mishima, K., Ushida, T. and Kikukawa, T. (2010) 512GB recording on 16-layer optical disc with Blu-Ray Disc based optics. *Proc. SPIE*, **7730**, 77300D-1–77300D-6.

[38] Usui, T. *et al.* (2010) *Investigation and Reduction of Signal Deterioration Caused by Submicron Vibration on Holographic Data Recording.* Technical Digest of International Symposium on Optical Memory. Hualien, Taiwan, October 2010.

[39] Park, S. and Hahn, J.W. (2009) Plasmonic data storage medium with metallic nano-aperture array embedded in dielectric material. *Opt. Soc. Am.*, **17** (22), 20203–20210.

[40] Wang, H. (2005) Studies of liquid crystal response time. PhD thesis. University of Central Florida, Orlando, FL, http://etd.fcla.edu/CF/CFE0000796/Wang_Haiying_200512_PhD.pdf (accessed January 5, 2016).

Index

Computational Liquid Crystal Photonics: Fundamentals, Modelling and Applications, First Edition.
Salah Obayya, Mohamed Farhat O. Hameed and Nihal F.F. Areed.
© 2016 John Wiley & Sons, Ltd. Published 2016 by John Wiley & Sons, Ltd.